Digital Transformation in Architecture and Construction

Edited by Alcínia Zita Sampaio

Published in London, United Kingdom

Digital Transformation in Architecture and Construction
http://dx.doi.org/10.5772/intechopen.1004677
Edited by Alcínia Zita Sampaio

Contributors
Alcínia Zita Sampaio, Buse Un, Claudia Szargan, Daniel Krause, Enes Yaşa, Ercan Erdis, Gülce Bozkurt, Kerry London, Maria Zuluaga, Selahattin Ersoy, Zelinna Pablo

First published in London, United Kingdom, 2025 by IntechOpen
IntechOpen is the global imprint of INTECHOPEN LIMITED, registered in England and Wales, registration number: 11086078, 167-169 Great Portland Street, London, W1W 5PF, United Kingdom

For EU product safety concerns: IN TECH d.o.o., Prolaz Marije Krucifikse Kozulić 3, 51000 Rijeka, Croatia, info@intechopen.com or visit our website at intechopen.com.

British Library Cataloguing-in-Publication Data
A catalogue record for this book is available from the British Library

Digital Transformation in Architecture and Construction
Edited by Alcínia Zita Sampaio
p. cm.

This title is part of the Civil Engineering Book Series, Volume 14
Topic: Construction Engineering
Series Editor: Assed Haddad
Topic Editor: Amin Akhnoukh

Print ISBN 978-0-85014-963-0
Online ISBN 978-0-85014-962-3
eBook (PDF) ISBN 978-0-85014-964-7
ISSN 3029-0287

If disposing of this product, please recycle the paper responsibly.

IntechOpen Book Series

Civil Engineering

Volume 14

Aims and Scope of the Series

Civil engineering is a traditional field of engineering from which most other branches of engineering have evolved. It comprises traditional sub-areas like transportation, structures, construction, geotechnics, water resources, and building materials. It also encompasses sustainability, risk, environment, and other concepts at its core. Historically, developments in civil engineering included traditional aspects of architecture and urban planning as well as practical applications from the construction industry. Most recently, many elements evolved from other fields of knowledge and topics like simulation, optimization, and decision science have been researched and applied to increase and evolve concepts and applications in this field. Civil engineering has evolved in the last years due to the demands of society in terms of the quality of its products, modern applications, official requirements, and cost and schedule restrictions. This series addresses real-life problems and applications of civil engineering and presents recent, cutting-edge research as well as traditional knowledge along with real-world examples of developments in the field.

Meet the Series Editor

Professor Assed N. Haddad is a Civil Engineer with a degree from the Federal University of Rio de Janeiro (UFRJ) earned in 1986, as well as a Juris Doctor degree from the Fluminense University Center earned in 1993, and a Master's degree in Civil Engineering from the Fluminense Federal University (UFF) obtained in 1992. He completed his Ph.D. in Production Engineering from COPPE / Federal University of Rio de Janeiro in 1996. Professor Haddad's academic pursuits have taken him to postdoctoral stays at the University of Florida, USA in 2006; at the Universitat Politècnica de Catalunya, Spain in 2010; and at the University of New South Wales Sydney, Australia in 2019. Currently, he serves as a Full Professor at the Federal University of Rio de Janeiro. He has held visiting professorships at various institutions including the University of Florida, Universitat Politècnica de Catalunya, Universitat Rovira i Virgili, and Western Sydney University. His research expertise encompasses Civil, Environmental, and Production Engineering, with a primary focus on the following topics: Construction Engineering and Management, Risk Management, and Life Cycle Assessment. He has been the recipient of research grants from the State of Rio de Janeiro, Brazil: CNE FAPERJ from 2019 to 2022 and from 2023 to 2025. Additionally, his research grants obtained from the Brazilian Government CNP since 2012 last to this date. Professor Haddad has been involved in several academic endeavors, being the Guest Editor of the International Journal of Construction Management; MDPI's Sustainability, Energies, and Infrastructures; Associate Editor at Frontiers in Built Environment / Sustainable Design and Construction; Guest Editor at Frontiers in Built Environment / Construction Management; and Academic Editor of the Journal of Engineering, Civil Engineering Section of Hindawi. He is currently a Professor of the Environmental Engineering Program at UFRJ and the Civil Engineering Program at UFF.

Meet the Volume Editor

Alcínia Zita Sampaio, Ph.D. in Civil Engineering from the Technical University of Lisbon, Portugal, has been an Assistant Professor in the Department of Civil Engineering at the same institution in Lisbon since 1999. Since 2001, she was the teacher and supervisor of Technical Drawing, Computer Aided Drawing and Building Information Modelling for Civil Engineering students. Since 2015, she has been a member of the Civil Engineering Research and Innovation for Sustainability (CERIS). Her main research focus is on the application of BIM and Virtual Reality (VR) in various construction sectors, including education and training, design coordination, construction planning, construction safety, bridge design, railway design and inspection, building maintenance, and Dynamo programming. She also supervises doctoral and master's theses related to BIM topics.

Contents

Preface

The construction industry has been admitting a strong change in the sense of introducing advanced digital technologies, modifying the traditional ways of working based on Computer-Aided Design (CAD) drawings, with evident benefits in its practical application and involving the various sectors of architecture and engineering. In this recent and necessary path, the BIM methodology has contributed intensely to the required digital transformation announced in the current well-known government directives, advocating for its rapid application with intensity and immersion throughout all sectors. The practical adoption of BIM has been accompanied by an open and varied interest in the field of academic research, with relevant results of great practical use, which have been reported in BIM scientific literature. The high level of digital transformation in construction, which is intended to be achieved in a complete way in all activities, contributes to streamlining the processes involved in the development of architectural and engineering projects for buildings and infrastructures, the planning of the construction process and the study of building maintenance strategies.

The degree of collaboration and integration promoted by the use of digital technology in the software available on the market has been growing rapidly in practical activity at all levels of design, real building work, and the use of the building. In accordance with the necessary dissemination of the benefits and limitations related to current digital technologies, namely Building Information Modelling (BIM), virtual reality (VR) or 3D scanner image capture, several research studies can be pointed out as innovative, in areas such as work health, safety design, project delivery, big open BIM, building production process and construction conflict management.

The introductory chapter points out the relevance of implementing BIM in the digital transformation of construction, involving all sectors, including architecture and the various engineering disciplines (Sampaio). Naturally, introducing advanced computer technologies promotes some distrust in the building community, which delays and hinders the penetration of BIM. However, the growing level of knowledge regarding the BIM concept and its wide applicability, among construction professionals, is either acquired in training new architects and engineers or in the evident verification of the benefits associated with its adoption in practice. As such, construction has been rapidly embracing BIM implementation, identifying a notable increase in the degree of collaboration, integration, and immersion that has contributed to improving the quality of the design process, the construction work, and the maintenance tasks.

In addition, London and Pablo consider the application of BIM in the health and safety construction sector. The study is focused on evaluating the aspect of work health and safety management tasks developed in a BIM environment, using the available BIM software. A comprehensive guideline was carried out, focused on aspects related to establishing the information requirements, the best practice definition considering the BIM adoption, and, finally, the tasks related to procurement, tendering, and supply chain monitoring. The study was elaborated together with the industry and government sectors, bringing a strong guarantee of its applicability in the construction domain.

Krause et al. refer to integrating a project delivery system with the interdisciplinary Big Open BIM, focused on the maintenance facility activity. A new integrated project delivery approach was considered and applied in the design and construction phases, fomenting added value that is supported by the detailed 4D BIM simulation, considering a correct scheduling of the work preparation and a digital cloud workflow-based collaboration. To achieve the required construction planning quality an integrated system approach using an open BIM standard was performed.

Virtual Reality (VR) technology can be linked to BIM, allowing users to improve the building production process (Bozkurt et al.). This research promotes improved control and fastening, imposing a systematic building construction process supported by the BIM+VR potential. Based on a literature review and a company survey, the authors identify the main strengths and weaknesses of this technological integration, as well as the evaluation of the potential threats and opportunities that this connection can bring.

Finally, Un and Erdis are focused on analysing the detection of construction conflict in a multidisciplinary project involving distinct stakeholders, using the most innovative technologies. The VOSviewer software was applied, allowing users to identify a strategic management of the construction conflicts collected in an early design stage. The study analysed the use of distinct innovative technologies in the construction conflict detection, namely, BIM, VR, digital rehearsals, computer vision, and the Internet of Things (IoT) sensors. From this, the integration of digital technologies presents a great potential to alert to the escalation of latent errors and risks into conflicts, claims, and disputes.

The distinct chapters of the book reveal relevant positive contributions considering the current application of distinct innovative digital software in the stages of the design development, the construction planning, scheduling and the conflict detection analyses. These chapters deeply disseminate the construction software's current progress and its application, showing how it can improve the construction quality of the final product.

The editor is deeply grateful for the relevant contributions to the constitution of the book, highlighting the current use of new digital technologies, to complete and add value to the different processes related to the construction industry.

Alcínia Zita Sampaio
Department of Civil Engineering,
IST, University of Lisbon,
Lisbon, Portugal

Chapter 1

Introductory Chapter: Role of BIM in Digital Transformation

Alcínia Zita Sampaio

1. Introduction

The adoption of Building Information Modeling (BIM) methodology in different sectors of the construction industry has admitted a strong governmental push, determined for public buildings and infrastructures, as a way to increase the level of digital transformation of the architecture, construction and operation (AEC) activities. This commitment has been followed, in each country around the world, from a horizon of 2007 in Finland to 2025 in Italy, passing through The Netherlands (2011), United Kingdom (2016), Spain (2018), Germany (2020), France (2022) and Portugal (2024). Observing government directives to increase digital transformation in the construction industry, the implementation of the BIM methodology has been introduced as the most adequate procedure, supported by advanced technologies capable of streamlining the development of integrated and collaborative projects, contributing to the achievement of accurate, sustained and efficient final products. Therefore, most professionals of the construction field have actively comprised BIM and implemented it in building projects that requires a multidisciplinary integration, in order to improve the performance of the large range of tasks and processes involved. Currently, BIM methodology has been used in complex building projects and as a research academic topic, reaching all the domains of the construction sector [1].

The BIM main concept considers the centralization of all information relating to the design and development of building projects, and involving complex integration of multidisciplinary design, construction and management of buildings and infrastructures. The BIM model, created for each project, establishes a mean of mitigating design errors, eliminating the duplication of information and reducing geometric inconsistencies between project phases. In it, the use of BIM platforms brings a height degree of integration and collaboration among sectors, tasks and teams contributing to optimize the final product.

The BIM systems, available on the market, combine several capabilities, such as the broad parametric concept (geometry, physical properties, identity and relationship rules), the friendly interaction aspect of the software in use, the ability of realistic visualization and the detection of conflicts between disciplines, supporting the elaboration of multidisciplinary projects. The high degree of integration and the efficient interoperability between software, provided by the advanced technological support of BIM work, afford a collaborative project with confidence in data transferred between partners, guaranteeing to reduce the repetition of information, to eliminate the eventual incorrect interpretation, and to agile the execution of the required project tasks. The implementation of BIM in the AEC enterprises, with all the valences enhanced by the methodology, requires a wide dissemination of its applicability and the recognition of an effective improvement of the working method, at an internal level.

The BIM project is prepared using advanced technology software, allowing the collaborative team to create, manipulate and add the information required along the development of a project, construction and management of a building. The created digital BIM model is structured according to the inherent BIM data organization, and it must present all the information always updated, following the distinct phases of the life cycle of the building. Based on this assumption, the 3D model, in each project, can support the development of nD BIM models, assisting the entire team involved in the building's life cycle.

1.1 BIM implementation: Benefits

The BIM methodology is currently the main work platform in the construction industry, and all sectors have been recognizing benefits in its adoption, enhanced by its multiple applicability with a high level of efficiency. The following aspects can be considered as the main benefits:

- The optimal collaboration in a team is based on the modeling process that depends of architectural, structural, mechanical, electrical and piping (MEP) adjustments, in order to conceive a complete and correct BIM model. The BIM software users identify, as the fundamental advantages, the easy modeling process based on the use of parametric objects, and the ability to integrate the development of all discipline solutions directly over the architectural and structural BIM components. In addition, the modeling procedure presents the possibility of individualizing each design component of the BIM model, an important capacity as it is carried out by a multidisciplinary team of architects and engineers. However, the correct integration of disciplines and the responsibility of the development of each component must be coordinated by a BIM manager;

- The data transfer process, frequently required to perform several type of analyses (structural and CO_2 emission) and simulations (energy and sustainability), is realized with a high level of accuracy. The BIM methodology supports the development of different components of the project, requiring the frequent transfer of models between systems, which must be supported by an adequate interoperability capacity between the specific BIM systems used. In a multidisciplinary design, the transfer process is normally well succeed. However, an important exception is verified in the context of the structural design, where the reinforcement bars are not transferred with sufficient correction. Despite this limitation, the modeling software contains easy-to-interact features that allow the user to correct and complement the reinforcement details [2];

- Over a correct and complete 3D BIM model, it is possible to manipulate the model database in order to support the development of technical drawings and the required nD models [3], such as 4D (construction planning), 5D (costs estimation), 6D (sustainability), 7D (maintenance and management), 8D (safety), 9D (lean construction) and 10D (industrialized construction) models;

- The Virtual Reality (VR) technology applied over the BIM model increases the potential of BIM in the construction sector, contributing to the achievement of a high level of collaboration in an immersive and interactive environment [4, 5].

1.2 BIM implementation: Limitations

Nevertheless, important limitations must be recognized in the BIM adoption in the design and in AEC enterprises. Two main topics can be referred:

- With regard to the design of structures, the deficient interoperability capacity, which is still verified among the available BIM base systems, constitutes a strong obstacle to the implementation of BIM in this domain. The most recent academic studies analyze the technological evolution of the modeling and structural calculation systems, observing the most problematic steps of the model transfer between systems. The technological advances that have been achieved have been increasing the functionalities available in BIM-based systems, used in the elaboration of multiple tasks inherent to the structural project. The degree of efficiency of the interoperability between modeling and calculation systems must be analyzed and known by the structural engineer, within the scope of the implementation of BIM in the office and in the project. Thus, the structural engineer must know the steps of the process, identify the limitations and develop strategies that lead to an efficient project.

- In the field of bridge design, the implementation of BIM presents some limitation due to the difficulty of defining appropriate and adjustable parametric objects for different concrete cases of bridges [6]. The generation of parametric models of bridge decks can be supported on the development of script using the visual programming Dynamo or the Python code, aimed to define new families of parametric objects, representative of the configuration of the bridge deck, with a rigorous definition of the cross section, longitudinal variation and geometry of the track layout. This is a complementary work that is required on the development of bridge projects.

2. Conclusions

The text summarizes the relevance of adopting BIM as the principal strategy to accelerate the digital transformation in the construction industry. As benefits, the data transfer processes made with efficiency, the possibility of manipulating the model database in order to obtain distinct nD BIM models and the improvement in integration and collaboration among the team involved in a design process, were carried out. However, there still are some limitations that must be known by the BIM users, supporting the definition a better BIM strategies.

Despite the limitations still verified during the development of BIM projects, prepared using the available platforms, the advantages pointed out outweigh the identified negative barriers. The adoption of the BIM methodology is currently the process that should be used in all sectors of the industry, contributing to its dissemination in various sectors and to the recognition of the evident benefits.

Author details

Alcínia Zita Sampaio
Department of Civil Engineering, University of Lisbon, Higher Technical Institute, Lisbon, Portugal

*Address all correspondence to: zita.sampaio@tecnico.ulisboa.pt

References

[1] Sampaio AZ. BIM multitask project manager: Responsibility, organization and interoperability. Procedia Computer Science. 2024;**239**:58-65. DOI: 10.1016/j.procs.2024.06.146

[2] Vilutienė T, Kalibatiene D, Hosseini MR, Pellicer E, Zavadskas E. Building information modeling (BIM) for structural engineering: A bibliometric analysis of the literature. Advances in Civil Engineering. 2019;**2019**:1-19. Article ID 5290690. DOI: 10.1155/2019/5290690

[3] Sampaio AZ, Gomes N, Gomes A. BIM design coordination: Conflict analysis and construction simulation. Procedia Computer Science. 2024;**239**:49-57. DOI: 10.1016/j.procs.2024.06.145

[4] Zaker R, Coloma E. Virtual reality-integrated workflow in BIM-enabled projects collaboration and design review: A case study. Visualization in Engineering. 2018;**6**(4):1-15. DOI: 10.1186/s40327-018-0065-6

[5] Sampaio AZ, Sarmento R, Gomes A. BIM improved with RV and AR technologies. Journal of Software Engineering and Applications. 2024;**17**(6):508-521. DOI: 10.4236/jsea.2024.176028

[6] Byun N, Han W, Kwon Y, Kang Y. Development of BIM-based bridge maintenance system considering maintenance data schema and information system. Sustainability. 2021;**13**(9):1-29. DOI: 10.3390/su13094858

Chapter 2

Co-design and Development of Building Information Modelling for Work Health and Safety Design, Construction and Management Industry Guidelines

Kerry London and Zelinna Pablo

Abstract

The construction industry has one of the highest high fatality and injury rates globally. Building Information Modelling (BIM) as an enabling technology can significantly eliminate and mitigate risks and improve work health and safety (WHS) management. Adoption is inconsistent, although various guidelines and legislation have been developed to reduce incidents and injuries in the workplace. Our literature review indicates that the UK, Singapore, Hong Kong, US, Germany, Spain and Finland public and private sectors are relatively advanced in the development of BIM technology. The US has been a significant contributor through its collaborative links with the UK, Australia, South Korea, Germany and Spain. In contrast, the integration of BIM and WHS management is less advanced. Australia, the site of this study has lacked research that evaluates WHS management in a BIM-environment. To increase BIM-WHS management integration across the industry, a government, industry and academic collaboration in Australia was undertaken, resulting in the development of comprehensive evidence-based guidelines comprising four key components: (1) Developing Information Requirements, (2) BIM for WHS Best Practice, (3) Procurement, Tendering and Supply Chain Monitoring and (4) Developing Project Information Requirements. The comprehensive iterative collaborative process that underpinned the development and distribution of the guidelines is described.

Keywords: knowledge transfer, information requirements, BIM WHS integration, data management standards, BIM and WHS management guidelines

1. Introduction

Construction is known as one of the most dangerous industries in which to work [1–4] and many safety incidents, injuries and fatalities could be prevented through improved design, planning and communication. Building Information Modelling

(BIM) is an enabler technology that involves the generation and management of digital Work Health and Safety (WHS) information in construction. BIM facilitates the separation of people and hazards through the use of technology and data [5].

The purpose of this chapter is to explain how industry guidelines for broadening adoption of BIM for WHS management were comprehensively codesigned and codeveloped with industry and government stakeholders as the key output from the research study. The overall aim of the study was to develop a better understanding of the application of BIM to support WHS management and, specifically, the role the client can play in enhancing adoption and application. Under this overarching aim, this chapter will explore the process for engaging industry and government throughout the study and the knowledge transfer methods to enhance development of useful and usable guidelines. The chapter presents the key outcome of a four phase study [6] conducted in New South Wales, a state in Australia:

1. Phase 1: Identify solutions for integrating the WHS aspect in BIM-enabled project planning, design and delivery.

2. Phase 2: Evaluation of WHS management in BIM-enabled project proposals.

3. Phase 3: Evaluation of the proposed approach for its adoption

4. Phase 4: Transfer of Knowledge and Dissemination.

It is noted that although the study has a local context the international setting is well considered and presented. The funder of the study is the multinational company Lendlease, a globally recognised enterprise as an innovator in digital construction, through the Workers Compensation Operational Fund. Additionally, the two key specialist subcontractors who provided much of the input into case study material for the Guidelines also operate in international markets. International experts in BIM for WHS management were interviewed; one from the UK and one from Singapore in Phase 1.

1.1 Background

There has been a plethora of research studies in the past decade that have analysed the adoption of BIM within the industry in an attempt to improve communication, information flow, efficiencies and productivity [6]. Equally, there has been numerous studies that have explored maturity and capability for adoption, identifying the need for new competencies and new models for adoption [7–13].

One major challenge of the construction industry is to find ways to reduce WHS incidents and injuries to support a safe workplace. A small but growing body of work suggests the use of BIM can facilitate the early identification of potential WHS issues and the application of preventative strategies using automated approaches. The strengths of using BIM for WHS management manifest in the embedded processes of visualisation, simulation, analytics, evaluation, and monitoring. It also supports communication, education and training during the design, construction, operational phases and post-construction phases. Documenting WHS requirements in plans and monitoring compliance with WHS legislation within a BIM environment can result in safer working conditions as WHS is identified as an integral part of the construction process, is made more visible to stakeholders, and can be appropriately managed at the various stages of construction [14].

Australia lacks research that evaluates WHS management in a BIM environment [15]. BIM-enabled WHS management is rarely considered in construction tender requirements and the evaluation of bidders' proposals, despite the scope for BIM to provide a powerful approach to the management of WHS. The management of WHS is likely to be most effective when considered holistically as part of the project life-cycle and included in a BIM-enabled environment by the construction procurer, with commensurate client response at the construction tender and evaluation stage. WHS requirements can be embedded early in the design and construction process [14, 15].

The UK, Singapore, Hong Kong and Finland public and private sectors are further advanced in the adoption of BIM and its use for WHS management. As research sites, they provided detailed information on the approaches, tools and outcomes highlighted by the theoretical, prototype and evaluative studies in this area [15].

Although not as advanced, the Australian private and public sector now provide sufficient examples of BIM-enabled construction projects that allow evaluation of BIM, from the tendering stage through to post-construction, and the potential creation of management and information systems for monitoring WHS. There also appears to be sufficient experience among industry and government stakeholders for major infrastructure projects to develop models for the adoption of BIM-enabled WHS management systems. Clearly, the brave new world of BIM holds potential for supporting the fulfilment of WHS obligations and it is timely to harness the potential of BIM to ensure safer construction sites.

This chapter presents the results of a study involving significant industry liaison and input. The study is part of a worldwide trend to develop better tools for collaborative decision making on construction projects through improved information ecosystems, including 'guidelines' to enable public and private sector client organisations who are key industry influencers to lead projects in a new way. The study explored the clients' role in catalysing a BIM-enabled WHS management ecosystem by incorporating it in procurement strategies and tendering requirements. This chapter is focused on the third phase of the four-phase study. Phase 3 of the study consisted of a comprehensive nine-step methodology (see details in **Figure 1**) to include iterative feedback from industry to develop, refine and analyse the potential impact of the content of the Guidelines.

1.2 Collaborative practice: Standards, guidelines and decision tools

To date, it appears that specific advice to clients on how to integrate WHS management in BIM has not been extensively considered in the procurement guidelines, particularly in Australia [15] Australian government agencies understand the importance of BIM for purposes such as efficiency and cost management. In partnership with industry organisations, government agencies have therefore developed extensive guidelines and procedures for its application and, through the government procurement process, are encouraging BIM development and adoption in order to harness its potential [16]. Having recognised the increasing benefits and potential for the wider application of BIM, Australian governments have in recent years actively supported its use and integration through the procurement of government-funded construction and infrastructure projects. For approximately 15 years, [16, 17] various government agencies and professional industry organisations have actively promoted BIM

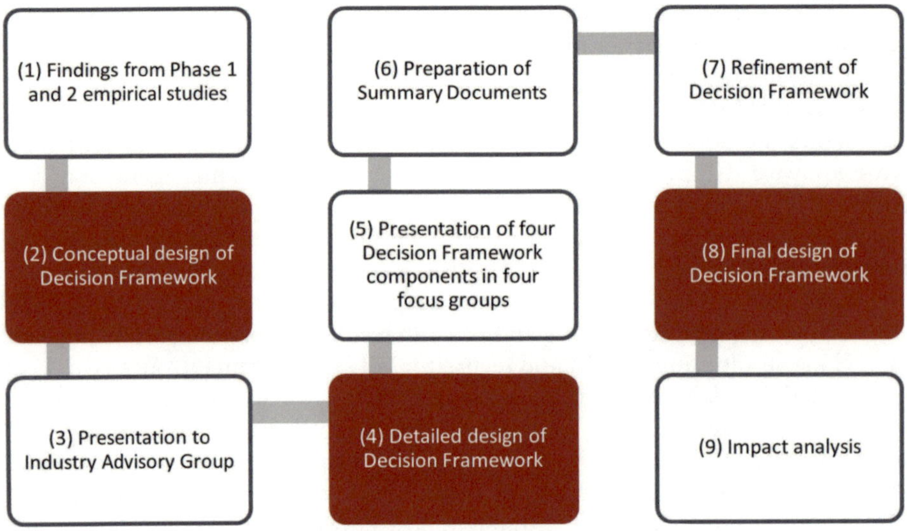

Figure 1.
Phase 3 method for developing the decision framework [15].

awareness and skills development for construction professionals to influence the uptake of BIM and its application across all stages of the project lifecycle.

Although there are numerous guidelines and tendering and procurement documents, there is little evidence of implementation and guidance on WHS management using BIM. There is no specific guidance on integration of BIM and WHS management, possibly because of the thinking may be that this may be too specific and prescriptive an approach to implement. Collaboration and sharing between project stakeholders of structured WHS information across the project lifecycle has been slow to evolve [14, 15].

As Australia moves towards adopting BIM for WHS management, the sector may gain insights from its international equivalents and adapt and modify to suit the local context. It is important to understand that while the most recent ISO 45001 sets out the WHS framework at international level, it is the standards, such as the UK PAS 1192-6, that outline the mechanisms by which the ISO 45001 health and safety management system may be applied using BIM [18–20].

There is currently no International Standard integrating BIM and WHS management although the ISO/FDIS 19650-6 Organisation and digitization of information about buildings and civil engineering works, including building information modelling (BIM) — Information management using building information modelling Part 6: Health and safety information is currently in development and planning approval phase. Once this standard is released it will have a significant impact on all construction sites globally with countries that are signatories to the International Standards Organisation.

Meanwhile the UK PAS 1192-6 is an exemplar that provides guidance on integrating WHS information through BIM processes and applications, ensuring safer work sites and reduced injuries and fatalities. PAS 1192-6 models a collaborative approach progressing through the project lifecycle between the client, designer, contractor, commissioner and end user. It places the onus on the client to clearly identify upfront the objectives of WHS requirements in a project. The UK EIR template has a section for Health and Safety and Construction (Design and Management) Regulations, which assists the client in defining how BIM will support WHS monitoring aligned with the different phases of the project. Details of how all related data can be captured and recorded is documented. Furthermore, the *BIM 4 health & safety* working group has outlined ten questions that a client should include in its EIRs to help prioritise the key WHS issues in line with the PAS 1192-6. The checklist is aimed at developing client understanding of how to integrate BIM and WHS management with the assumption that it is easier to assess WHS management in a BIM-enabled environment when it is integrated in the tender documents in a holistic manner rather than when contained within standalone documents [19, 20]. All in all, there is potential for incorporating BIM enabled WHS management at the procurement stage as clients can more holistically and effectively identify, articulate and ensure WHS integration in the construction lifecycle and workplace.

2. Development of decision framework methodology

The purpose of this study was to develop a Decision Framework that would provide industry guidelines supporting the development of a digital data and information management ecosystem that synthesises BIM and WHS management across the entire life cycle. The Decision Framework aims to move from the generic approach of BIM adoption to a targeted approach as BIM as an enabler of achievement of a project objective (i.e., safety) with specific exemplars, case studies and tools orientated towards that objective. The Decision Framework provides guidance on how to develop an Organisation Information Requirements document that supports the Project Information Requirements document, particularly in enabling the development of high-quality criteria for tendering, evaluation and monitoring, and ultimately, the delivery performance of the primary contract throughout the supply chain.

To develop the Decision Framework, case studies, document analysis and semi-structured interviews were conducted. BIM and WHS management toolkits and resources were also examined [19, 20], including those provided by the UK-based BIM4WHS Working Group [18]. Guidelines and standards such as the NSW Infrastructure Development Management Framework [21] and ISO 19650 [22] have also provided a much-needed context to this potentially challenging task of bringing two worlds together of "BIM" and "WHS". The following diagram summarises the nine-step process undertaken in Phase 2 (Steps 1–3) and Phase 3 (Step 4 onwards).

Figure 1 indicates that the BIM for WHS Management Decision Framework was built from the synthesis of the outcomes of Phase 1, i.e., desktop literature review and preliminary interviews with experts, and Phase 2, an in-depth empirical study which resulted in the recommendation of guidelines with four components, the purpose of each component and the principles for development in Phase 3. The process illustrated in **Figure 1** indicates that the Decision Framework developed by the research team went through three versions (conceptual, detailed and final design), was underpinned by empirical studies and was subject to various levels of validation by industry experts. The levels of validation are presented in greater detail in **Table 1**. The BIM for WHS Management Decision Framework is structured in a hierarchy as indicated in **Figure 2**. The Framework is composed of two sets of guidelines: Information Requirements guidelines and Project Information Requirements guidelines.

The Phase 2 interviews, analysis and themes provided significant input to the Decision Framework text for case studies and the various sections. As discussed in Phase 2 Technical Reports [15], the proposed Decision Framework supports pathways

	Component 1A	Component 1B	Component 2A	Component 2B
Phase 2 empirical study: project interviews	✓	✓	✓	✓
Phase 2 empirical study: client interviews	✓	✓	✓	✓
IAG Workshop:	✓	✓	✓	✓
Focus Group 1:		✓		
Focus Group 2:			✓	
Focus Group 3:	✓			
Focus Group 4:				✓
Additional interview for more material; 2 case studies		✓	✓	
Direct editorial comments	✓	✓		✓
Case study impact analysis	✓	✓	✓	✓

Table 1.
Levels of validation.

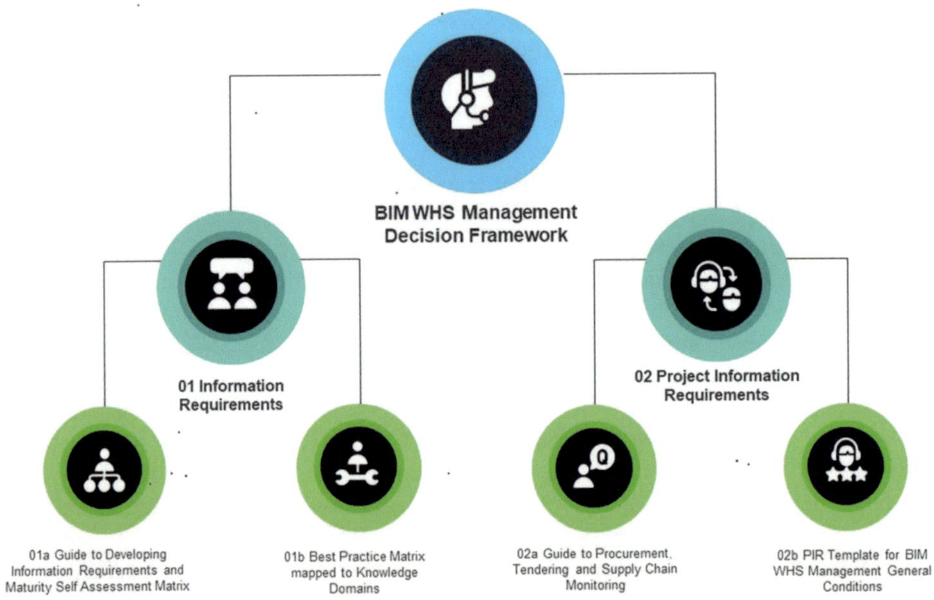

Figure 2.
Hierarchy of decision framework guidelines [15].

for clients and project team leaders to collaboratively work together and develop high-quality information requirements that clarify the data management environment expectations for all stakeholders at an organisational, asset and project level. **Figure 3** summarises the development, testing and validation of the guidelines throughout the study and the relationship between Phase 1, 2 and 3 activities and the findings.

Components 1 and 2 are companion documents that include guidelines. The Information Requirements (Component 1) provides a strategic discussion for public

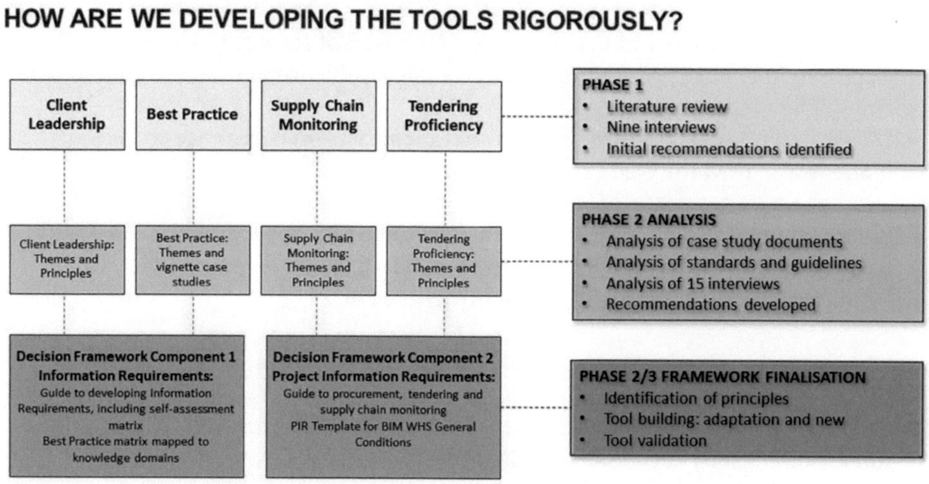

Figure 3.
Integration phase 1, 2 and 3 key themes and activities for decision framework development [15].

and private sector clients to establish themselves as leaders in establishing the planning, designing, constructing and managing assets lifecycle in a safe manner. The Information Requirements align with the strategic approach as outlined in NSW Infrastructure Development Management Framework and it also follows an international trend for integration of BIM (digital engineering/digital twinning) to improve project management and project performance. Concepts such as Organisational Information Requirements and Project Information Requirements are explained and a practical case study of how to develop such documents is provided based on the international standards. A Self-Assessment Maturity Matrix is also introduced aimed at early adopters, leaders and innovators, specifically targeting BIM and WHS integration. Best practice case studies are also provided, drawing from both the international literature and empirical fieldwork conducted in NSW; specifically, four best practice case studies from Australia and ten international exemplars are presented that give practical examples to organisations of what is possible. The inter-relationships between a project lifecycle, the information requirements and models, the knowledge domains and the level of development of a model are explained to give guidance to the Decision Framework users. The Guideline documents can be found on various websites [23].

2.1 Detailed co design and development methods

This section provides an overview of the approach taken to develop, refine and analyse potential impacts of the contents of the Decision Framework. The following **Table 1** maps each component to the research activity and the following sections summarise each activity.

2.2 Step 1,2 and 3 conceptual design

An in-depth discussion of the conceptual design development of the Decision Framework can be found in the Phase 1 and Phase 2 Technical Reports [6, 15]. Briefly, Technical Report 1 summarised findings from a targeted literature review that was subsequently validated through an empirical study comprising nine interviews. Findings highlighted four key drivers of BIM adoption: Client Leadership, Tendering Proficiency, Supply Chain Monitoring and Best Practice. In Phase 2 of the study, the four drivers were explored in extensive detail through an analysis of a broad range of data sources:

- 13 semi-structured interviews with participants from three case studies that made use of BIM for WHS management
- two semi-structured interviews with directors from government clients with leadership in BIM for WHS management, which had a direct bearing on the framework design and are discussed further shortly
- 50 project documents
- existing BIM and WHS management toolkits and resources [19, 20], including those provided by the UK-based BIM4WHS Working Group [18]; and
- guidelines and standards such as the NSW Infrastructure Development Management Framework [21] and ISO 19650 [22].

Following Phases 1 and 2, the four recommendation areas identified were mobilised as the main scaffolds for the Decision Framework. The early design comprised of two key components and four subcomponents linked in a hierarchy (shown earlier as **Figure 2**). An important feature of the Decision Framework, even at its early stages, was that it was already significantly aligned with the NSW Infrastructure Data Management Framework [21], particularly in relation to terminology, agency information requirements and the types of models and their interrelationships. The IDMF further aligns with the ISO 19650, making the content of the Decision Framework not only relevant to the NSW context but also relevant internationally.

Once high-level conceptual designs for the four components were developed, the research team organised a feedback session with the Industry Advisory Group (IAG), a group of professionals from government, academe and industry. The IAG members collectively had expertise in construction, design, WHS management and specialised areas like craneage and BIM. The IAG was established at the very start of the project and was convened every 3 months. Research team members regularly presented findings and obtained systematic feedback during these quarterly meetings. Key feedback sessions were recorded and discussions transcribed. Members of the research team then incorporated input and suggestions that arose from these discussions as the project progressed.

At the time of presentation to the IAG, the basic structures and general content descriptions of the components had been established. However, detailed contents had yet to be finalised and proposed component outlines were still fluid and open to discussion. For example, Component 1B was envisioned by the research team to be a Best Practice Matrix in some form. During this initial presentation to the IAG, the Best Practice Matrix contents were presented as high level concepts. This strategy, of presenting open-ended and flexible proposals, was intentionally selected to give room for nuanced input. The final version of this matrix, contained descriptions of six Knowledge Domain areas for BIM for WHS management, four Australian case studies, international case studies, visuals and links to sample videos.

Feedback was obtained from the IAG through a 1.5-hour focus group. A packet of materials, including the conceptual designs of the four components and guide questions, was distributed 1 week prior to the session. Six members of the IAG participated in the focus group (refer to **Table 2**). The research lead opened the session with a preliminary presentation of the components. After this initial presentation, small breakout sessions comprising two IAG members and two or three members of the research team took place. Three guide questions were provided; these were intentionally framed in broad terms to elicit wide-ranging input. Discussions were led by IAG members themselves; the eight research team members that were present asked questions, took detailed notes and made clarifications. After the breakout sessions, all participants reconvened to consolidate feedback and for inter-group sharing. The discussions were recorded for each group, then consolidated into a master list.

Following the IAG feedback session, two industry participants also provided written feedback through comments and changes on editable drafts using track changes.

2.3 Step 4–6 framework detailed design

The presentation of the Decision Framework's conceptual design to the IAG marked the close of Phase 2 of the research project. Phase 3 then commenced with team members being deployed to sub-teams for detailed development of the four components. The project leader developed detailed structures for each of the four components (refer to **Table 3**). At this stage, detailed development was driven by empirical findings (described in Step 1) and by the IAG feedback.

Following the development of the components' detailed design, the research team organised four workshops as contexts for focus group interviews. The aim of the workshops was to gather representatives from government agencies, general contractor companies and specialist contractor companies to provide expert stakeholder feedback on, and analysis of, each of the components. Unlike the earlier IAG

Group 1	• Multinational 1 WHS expert Chair • SME 1 BIM expert • Academic researcher 1 Institution 1 Note taker • Academic researcher 2 Institution 2 Note taker
Group 2	• Multinational 1 WHS expert Chair • SME 2 BIM expert • State government representative • Academic researcher 3 Institution 2 Note taker • Academic researcher 4 from government organisation (SW) Note Taker
Group 3	• Multinational 2 BIM expert) Chair • Multinational 1 BIM expert • Academic researcher 1 Institution 2 Note taker • Academic researcher Institution 2 Note taker • Academic researcher Institution 1 Note taker

Adapted from Ref. [24].

Table 2.
Participants in IAG feedback session and groupings.

Component	Key sections identified at the start of detailed design	Other remarks/parameters
Component 1A: Guide to Developing Information Requirements with Self-Assessment Matrix (47 pages)	• Purpose and users of the Guide Note • Discussion on integration of information requirements, knowledge domains and BIM LODs • Overview of leadership and collaborative practice • Sample case study showing OIR, PIR and EIR integration and samples for logistics planning • Self-Assessment Matrix comprising 10 questions developed based on Phase 1 and 2 findings	• Resource is significantly novel • Some sections loosely based on selected parts of BIM Alliance's Guidance Note D; however, discussions are strategic and are empirically-informed • Case study on OIR, PIR and EIR for logistics planning takes a coherent supply chain perspective • Self-Assessment Matrix was conceived based on BIM4WHS Working Group's Plain Language Questions for Developing Employer Information Requirements. However, Employer Information Requirements are no longer as central given the publication of ISO 19650, so the Self-Assessment Matrix offered here is significantly different, with a more strategic OIR flavour
Component 1B: Best Practice Matrix (5 A3 pages)	• Six knowledge domains for BIM for WHS management • Australian case studies mapped to knowledge domains • International case studies mapped to knowledge domains	• Resource is new; no similar existing resources currently available
Component 2A: Guide Note to Procurement, Tendering and Supply Chain Monitoring (40 pages)	• Purpose and users of the Guide Note • Procurement strategies and their role in BIM for WHS management • Case study on complex timber façade • Guidance for pre-tendering, tendering and post-tendering with checklists • Guidance for supply chain monitoring with checklist	• Resource is significantly novel • Some sections loosely based on selected parts of BIM Alliance's Guidance Note D; however, most sections (checklists, case studies) are new and are supported by empirical analysis
Component 2B (4 A3 pages)	• PIR General conditions • PIR Specific conditions for lifting	• Conditions were based on BIM4WHS Working Group's template of general and specific conditions) a work in progress) • General conditions have been categorised and the language reframed to be less prescriptive and more suited to the Australian market • Specific conditions were developed based on empirical data and validated by industry professionals who were part of the IAG • Selection of "lifting" was driven by the large volume of empirical data that came from the case studies

Table 3.
Key features of components [24].

session where participants provided feedback on the high-level conceptual design of all four components, these four workshops were aimed at eliciting in-depth feedback on the detailed design of a single component. In each workshop, participants were expected to qualitatively assess the feasibility of the Decision Framework as the recommended solution for enabling significant adoption of BIM for WHS management for the sector. Participants for each focus group were selected based on the nature of the contents of the component to be discussed. For the focus group on Component 1A, for example, participants invited were professionals connected with government clients, because of the strategic nature of the document and its emphasis on high level information management decisions. For Component 2B, which had a more technical, project-based focus, participants included construction as well as health and safety specialists, with validation by a crane specialist for the "complex lifting" specific conditions. A list of participants in each focus group is provided in **Table 4**.

The research team did not find it necessary to conduct focus groups sequentially. A decision was made instead to begin with Component 1B, which had a clear matrix structure and could thus be easily modified. Prior to each focus group, participants were provided with (1) a draft of the component; (2) an information packet with background on the research and (3) a list of questions (refer to **Table 5**).

Focus group discussions were documented by research team members, who again cross-checked and then consolidated their notes. Focus group feedback was then transferred into Summary Documents, where they were combined with comments from client interviews and from an IAG. Summary Documents thus served as comprehensive repositories of revisions for the components.

Focus group topic and date	Participants/affiliation
FG 1: Component 1B	• Director, The Men from Marrs (with additional meeting for a fourth case study) • Head of Environment, Health & Safety, Building, Lendlease (written feedback also obtained) • Head of Digital Engineering, LendLease Integrated Solutions (written feedback also obtained)
FG 2: Component 2A	• Head of Design & Project Management (NSW) Lendlease • Construction Manager, Lendlease • Head of Client & Strategy NSW & ACT, CPB Contractors
FG 3: Component 1A	• Project Officer, Office of Projects Victoria • Project Officer, Office of Projects Victoria • Senior Manager, Office of Projects Victoria (written feedback sent) • Director, Testsafe and Engineering, Safework NSW (met separately on 6 July 2021, written feedback also obtained)
FG 4: Component 2B	• Head of Client & Strategy NSW & ACT, CPB Contractors • Head of Environment, Health & Safety, Building, Lendlease (written feedback also obtained)

Table 4.
Focus group details.

Question	Possible prompts
What is your feedback on the Guide on Developing Information Requirements?	• How well does the Guide meet the purpose as explained in previous session(s)? • How can it best be used by your organisation? • Which roles is it most useful for? • What are some comments you have on each of the main sections? • What are your comments on elements like: language, layout, use of visuals, density, overall appearance, length?
What would you change about the components and why?	• If you wanted to change anything, how would you change: language, layout, use of visuals, density, overall appearance, length? • For content: What would you add, and why? • For content: What would you remove, and why?
What would you recommend to make the Component more generally applicable to a range of players, including SMEs?	• Does the tool exclude anyone? • Could you identify ways that it meets/ does not meet the needs of major players vs. smaller businesses?
What other comments do you have?	• (free discussion)

Table 5.
Focus group structure and questions for Component 1A [24].

Focus groups were recorded and discussions transcribed. Sample feedback for each component is presented (refer to **Table 6**). All feedback from a focus group was captured in a Summary Document. The complete Summary Documents are discussed further in the results section in the Technical Report No 3 [24].

2.4 Step 7, 8 and 9: Final design refinement of decision framework

Feedback was categorised into three groups: confirmation-related, content-related and editorial-related in the Summary Documents. Components in this step would be revised, leading to final design. Feedback captured in the Summary Documents did not suggest major changes in any of the components. Many comments validated proposed contents. Changes proposed usually involved enhancing material that was already available. Participants did not suggest that any element or section be removed.

Following final design, the research team undertook one more step in relation to the Decision Framework's potential benefit or contribution. This step involved an impact analysis wherein the four components were considered as analytical devices to assess information requirements of two past projects. During the research co-design stage, it was established that projects selected ("case studies") for this stage could either be (1) a BIM-enabled project with potential of WHS Management aspects or (2) a past project that included WHS Management aspects but was not a BIM-enabled project. The aim of this impact analysis was to critique the actual project requirements of real cases (OIR, PIR and EIR) against the ideals of the Decision Framework. The framework's ideals were distilled into a case study protocol comprising 26 questions, which were then used to conduct a desktop analysis. Details on the protocol and its use are provided in the Discussion section. The protocol was applied to two case studies, one involving the design and construction of an underground metropolitan train station, pedestrian station access and two retail and residential towers and the second

Component 1A Guide to Developing Information Requirements	1. A strong niche that merges BIM and WHS management has been identified. 2. Consider providing more context in the beginning, by noting that this is work on BIM then moving on to how BIM for WHS management is the gap that the team is trying to fill. 3. Add a lead in on the importance of information management.
Component 1B Best Practice Matrix	1. The matrix is already inclusive. The practices shared can be done by large as well as small and medium firms as long as management drives the use BIM for WHS management; size does not determine this. 2. Models need to be designed for purpose. Models aren't generally useful for purpose and need to be designed to a standard and show how to do things and basic checks. 3. Need to work towards developing standardised models that are computer literate (quantification) to avoid non-use, complexity and cost and encourage standard interpretation.
Component 2A Guidance Note on Procurement, Tendering and Supply Chain Monitoring 48 items	1. Perhaps the integration could be brought into even sharper focus in the first section by introducing a table/diagram that maps LOD/4D Modelling with WHS Management phases in the Introduction section in the Client part at the beginning. 4D – sequencing, 5D – cost, 6D – life cycle. And link this to a risk mgt process 2. Consider a different case study example specifically on WHS and with more 3D images 3. Suggest a second example/case study. Perhaps a study where things did not work so well and interventions and corrections were made to get back on track.
Component 2B PIR General and Specific Conditions	1. PIR template – General: Tag to existing Australian or international standards to streamline the number of standards that projects have to benchmark against it 2. PIR template - For the specific areas: Cranes can be considered as one of the specific areas since they are a huge safety risk

Table 6.
Sample feedback on each component adapted from Ref. [24].

involving the design and construction of a dual-carriage motorway in Sydney's west. Case Study 1 was one of three analysed in Phase 2 of the research while Case Study 2 was a new case analysis undertaken during Phase 3. For both case studies, the research team made a list of documents to request for review, using online client information on OIR, PIR and EIR. For Case Study 2, documents provided included

- A quality specification document
- A digital engineering execution plan, submitted as a response by the main design contractor
- An excel file showing BIM specifications and schemas for the project

One constructability workshop report.

For Case Study 2, interviews were also conducted with the project team's Technical Director/Digital Engineering lead as well as with its Digital Engineering Manager.

3. Results

The results section has two major sections including (1) Feedback and (2) Decision Framework. As discussed in the Methodology, the Feedback section draws from various sources and these recommendations are compiled in four Summary Documents [24]. The Decision Framework section includes the final four components which are a key outcome of this study can be found in various website locations including Torrens University Website Featured Research [25] and Centre for Work Health and Safety [26].

3.1 Feedback

Feedback on the components were obtained from client interviews which were conducted in Phase 2, the IAG feedback session and four focus group discussions. Written feedback was also obtained from specific participants following focus groups.

3.1.1 Feedback from client interviews (phase 2)

As noted earlier, the research team began gathering preliminary feedback as early as Phase 2 of the research. The earliest source of feedback was client interviews, which were conducted in the during the case study analysis and well before any high-level design of the components had been developed. At the time of the interviews, only some of the contents of the framework had been decided upon. For example, the research team had already established that a Self-Assessment Matrix in some form was essential but had not yet established its contents. The team's method of obtaining feedback involved showing interviewees existing toolkits developed by other bodies such as the BIM4WHS Working Group. Questions were raised as to how these existing tools could be improved, adapted, or even significantly expanded to suit an Australian context.

Interviewees were also asked to clarify some open issues. For example, the research team had very productive discussions with client interviewees about their views on the significance of different procurement models in influencing the effectiveness of BIM and WHS management integration. These discussions later influenced the contents of Component 2A, even if at the time of client interviews, Component 2A's structure had not yet been defined. Examples of feedback from the client interviews, specifically recommendations about the Self-Assessment Matrix, are summarised (refer to **Table 7**).

All comments from the client interviews were taken into consideration in the design of the components.

3.1.2 Feedback from the industry advisory group

Following the client interviews in Phase 2, the research team developed the conceptual design of the four components and presented to the IAG. Sample IAG feedback is provided in **Table 8**.

3.1.3 Feedback from focus group interviews 1,2,3 and 4

Following the presentation of the conceptual design, the research team developed the detailed design of the Decision Framework. Each of the four components was then presented in a focus group. In total, the number of comments received for each component from three rounds of feedback was as follows:

Client interviewee 1	The value of the current UK format [the BIM4WHS' version of the maturity matrix] is unclear; it bounces between digital (CDE) with non-digital (safety management).
	Step One [of the matrix] should involve examining existing WHS processes, noting that these can be very project-specific.
	Step Two should then focus on understanding the data needed to support these processes. Technologies are important but are secondary to understanding the importance of data.
	Step Three (?) should focus building a data "ecology" for the project. Questions must be asked to bring others along on the journey and create an integrated digital ecosystem, with the digital engineering developer bridging the disconnect between BIM and WHS
Client interviewee 2	Maturity levels in the matrix are useful because government clients will have projects that vary in their safety emphasis.
	Language should be more clearly linked to BIM and should be modified so that it "resonates"
	Consider that there are even moves in the UK and in Victoria to interrogate the use of the term "BIM"
	Include question on process for hazard identification
	Include question on scope of multi-stakeholder participation
	Language should be plain English but also relevant
	Include question on having a plan in place to construct assets
	Include question on access to BIM model

Table 7.
Sample feedback from client interviews about the self-assessment matrix.

Group 1	• All components are valid • Component 1a: The guide needs to cover high-risk activities • Component 1b: Knowledge domains are valid. It would be good to consider mentioning other sophisticated digital tools outside of BIM, for example algorithms that can help correlate designs and falls from height
Group 2	• Component 1a: The guide should put safety as the focus of every stage of the process. The guide should consider contractor capability or at least ways to bring subcontractors up to speed • Members confirmed that the self-assessment matrix is very important, particularly at the beginning of a project • Component 2b: Supply chain monitoring guide is very important. The guide should use specific instructions as these are easier to follow than high-level principles. For communication, examples of implementation of procurement and tendering best practice (cases, scenarios etc.) need to be provided
Group 3	• Component 1a: Focus on 4D as a priority • Component 1b: Happy with the Knowledge Domains. It would be good to have a small-scale project in a large-scale project example for each of those domains so that they can be tangible to clients and the size of the project that they undertake • Happy with Component 2a and 2b • Self-assessment matrix is a good strategy • Communicate through industry associations like Engineers Australia/MBA/Build Smart/PCA

Table 8.
Sample feedback from presentation of conceptual design to IAG.

- Component 1A (Guide to Developing Information Requirements): 52 comments
- Component 1B (Best Practice Matrix): 108 comments
- Component 2A (Guide Note for Procurement, Tendering and Supply Chain Monitoring): 48 comments
- Component 2B (PIR General and Specific Conditions): 38 comments

Participant comments could be grouped into three areas including: (1) confirmation-related, where participants agreed with our work, (2) editorial-related and (3) content-related. The first set of comments involved participants' validation of contents of the components. Examples of such comments were "The matrix is already inclusive" and "A strong niche that merges BIM and WHS management". The Summary Document of Component 2A, for example, had 48 comments; 23 of these were "no action" items because participants reaffirmed the content.

A second set of comments involved editorial changes, or changes in terminology, language or tone. Suggested changes mainly involved reshaping the text to better suit the Australian industry. In one focus group, for example, participants raised concerns around terms such as "surveillance" and suggested the use of the alternative "assurance". In other situations, suggestions were made to keep wording more "neutral". For three out of the four groups (FG1, FG3 and FG4) the research team sent out drafts to selected participants, who provided their editorial suggestions in track changes on the document itself. In FG2, there were very few editorial comments, so this step of direct editing was not deemed necessary.

A third set of comments involved changes in content. Most of the comments in this category were suggestions about adding more resources to supplement existing content, so that examples and case studies could be presented in ways that were more vivid. For example, participants suggested that the team provide more visuals showing 4D BIM more case studies and additional checklists. Almost all suggestions in this group of comments were incorporated. In some cases, suggestions for one component were found to have been addressed through the contents of another component. For example, one suggestion about Component 2A was to include a case "where things didn't work so well". The research team found that this suggestion was addressed in Component 1B, the Best Practice Matrix, which showed case studies that had clear gaps in the use of BIM for WHS management, gaps that signalled missed opportunities to use BIM to support specific Knowledge Domains. The most significant suggested addition related to Component 2A involved two participants suggesting the need for a high-level diagram that consolidated key dimensions of the project.

In very rare cases, a few suggestions were not implemented, or only partially implemented, because they were beyond the research scope. One participant in the Component 1A workshop, for example, suggested that the Guide to Developing Information Requirements include detailed implementation guidelines on how to develop a universal language, the way that SMARTBuild or IFC had been seeking to do. This suggestion falls outside the scope of Component 1B. Component 1B was designed as a high-level document providing guidance for developing strategic-level Organisational Information Requirements (OIR). The team responded by writing a section with a short, strategic discussion about "Data Ecosystems and Universal Data" but did not draft detailed operational guidelines. It is noted, however, that the suggestion was an important one because standardisation of data is seen to be a fundamental requirement for any collaborative approach using BIM. This topic is a high priority area of future research. The following sections provide details on the outcomes of each of the focus groups.

The research team made a decision that the first focus group discussion would focus on obtaining input on Component 1B: Best Practice Matrix, which was seen as a practical starting point because of its discrete, identifiable elements and its simple structure. Component 1B's building blocks were local and international case studies that were analysed using the six Knowledge Domains. The table format of matrix made it a feasible 'jump-off point' as compared to other more narrative-based documents [24].

Key content-related themes that emerged from the first focus group were minor and these included: putting in a fourth local case study, with one case on craneage being offered by one participant; replacing one international case study which involved automated remote monitoring, with one focus group participant noting the case could be perceived as controversial in Australia and including additional resources such as video links and visuals. Editorial comments were also straightforward. Other comments involved suggestions for better consistency in the use of some terms and suggestions for the replacement of other terms such as "surveillance". Sample comments about Component 1B are shown (refer to **Table 9**).

Comments: confirmation-related	
Comment	**Response**
Great example and application. Solving in a safe and virtual space first is one of the key benefits of BIM and Digital Engineering of EHS.	Agreed. No action required.
Great example of increasing collaboration. EHS is often the "Trojan Horse" to get other beneficial practices adopted in AEC, so increased collaboration and emphasis on solving issues in the virtual/design space is fantastic. Here an opportunity exists to code some of the NSW/AU EHS requirements into BIM as a plug and play.	Agreed. No action required.
Add potential case study of removal of cranes from Crown Casino. We used the model to develop a methodology around removal of cranes.	Agreed. Case study 4, which focuses on cranage, has been added following an interview with Simon Marr.
Remove the final example. It is controversial, would not work in Australia. It could cause the shutdown of the site. Privacy issues, liability issues etc. Australia is not ready. Expensive. An alternative would be to use computer vision of safety zones and instead of real-time warnings, use the data to discuss breaches of safety zones at the end-of-day safety review.	Agreed. This case study has been removed and now replaced by a new case study on 4D BIM based workspace planning for temporary safety facilities
Comments: editorial-related	
Matrix language may be threatening to industry and cause push-back. See matrix oversees road project, with words such as 'surveillance' recommended to be replaced by 'assurance'. Make wording neutral.	Agreed. "Surveillance" is now replaced by "assurance".
Matrix language could be looked at and you are going to provide us tips ...ensure consistency of language e.g., risks and hazards; surveillance and assurance.	Agreed. R Trethewy's editorial comments have been incorporated.
BIM facilitates the separation of people and hazards using technology and data - misleading comment: it facilitates identification of hazards and how to plan for possible separation. BIM in and of itself does not do that.	Agreed. Changed wording as noted previously.

Table 9.
Partial summary document-component 1B.

Two participants had written feedback prepared before the focus group. These were obtained by the research team. These written changes generally involved editing for sharper technical precision (refer to **Figure 4**). Examples of changes are in lighter text.

A final point to note is that a total of 108 comments were received during this focus group, more comments than in succeeding focus groups. This number must not be misconstrued to mean Component 1B was of lower quality. As noted, most of these comments were minor. Second, Component 1B was the first of the components to be assessed, and several suggestions that were received were actually meant for future focus groups involving the other components. A third point is that at the time of the first focus group, many aspects of the other components had not yet been finalised, so discussions in this focus group had less restrictions and parameters. The open-endedness of the session understandably led to more exploration and suggestions, since team members were not yet constrained by the later decisions that would rule out sets of options.

The second focus group's aim was to gather feedback on Component 2A: Guidance Note for Procurement, Tendering and Supply Chain Monitoring. As with the first focus group, participants confirmed the value of many aspects of the component, including the importance of the case study and actionable items like the checklists. Editorial-related comments were minimal, including the need to specify "construction" actors as users of the report and the need to ensure that section headings provided signposts that supported a clear narrative thread. After the focus group, the research team made the decision that it was not necessary to send out written drafts for direct editing. Instead, the team focused on addressing the mainly content-related comments. Based on feedback, an additional case study on early contractor involvement was written; existing checklists were expanded a new reverse checklist supporting main contractors in their engagement of subcontractors was developed and new visuals were also added to break up the more extended narratives. Two high-level diagrams were also added [24]. The first diagram was a conceptual integration of project stages, Knowledge Domains and LOD; a second was a hierarchy that visually depicted how the four components were linked to one another.

CASES

Description	Scenario Planning	Requirement Briefing
ASSET: A complex six-storey circular commercial building including a facade that was wrapped in 20 kilometres of sustainably-sourced timber strips. **CLIENT**: A multinational representing the developer, design manager and construction manager, with each function executed by a separate division. **KEY BIM DRIVER:** An independent architectural consulting firm completed the concept design. The facade subcontractor was the primary catalyst for the use of BIM because of the complexity of the circular design and its façade which included externally fixed timber strips wrapped around the structure. The facade subcontractor manufactured the facade components offsite and transported them to a inner-city site for installation, which was surrounded by public domain. Careful preplanning and logistics were critical to the successful integration. BIM was key to ensure detailed design elements were visualised which included safe transportation, unloading and installation of the façade and other key design elements.	**USE CASE**: The Model was used for design planning to determine the optimum location for cranes used to install the facade. **BENEFIT:** The facade subcontractor identified critical design angles and dimensions within BIM to ensure that cranes could be positioned correctly the first time to avoid rework and related safety, time and cost implications.	**USE CASE**: The Model was used for planning detailed design and related work sequencing. Daily work activities were then shared with supervisors on site. **BENEFIT**: The facade subcontractor provided BIM-images of daily work activities, which were used by construction supervisors to brief other trade subcontractors on the site during toolbox talks about WHS management like the location and extent of exclusion zones and high risk activities, such as slinging and lifting.

Figure 4.
Sample editorial changes made in component 1B. Note: Table with participant changes in grey.

During the third focus group, participants were asked to provide feedback on Component 1A: Guide to Developing Information Requirements. Component 1A, which focused on high-level discussions on matters such as supply chain culture, long-term goals and Organisational Information Requirements (OIR), was the most strategic of all the components. A decision was thus made to invite government client representatives to the discussion as they were familiar with high-level leadership issues such as these. Three representatives from the Victorian state government agreed to provide feedback, along with a fourth participant from the NSW work health and safety regulator, who was met separately. An important consideration about the three representatives from Victoria was that they were not part of the IAG. This meant that these participants were hearing about the project for the first time and had limited background information on the component they were being asked to assess. Focus Group 3 therefore provided the research team with an opportunity to assess if too much prior knowledge had been assumed in the development of the document. The Victoria-based participants confirmed that the overall goal of the research, of integrating BIM and WHS management, was an important and productive niche and that the research was very relevant. They sought clarification on certain sections, for example what the purpose of the Self-Assessment Matrix was and how it was to be used.

The discussion led to a number of changes. For example, the research team decided to prepare new sections about the Self-Assessment Matrix, clarifying that it was not a quantitative device for benchmarking against the rest of the industry. A brief explanation was developed to note that it was an internal document and conversation-starter which was meant to encourage supply chains to move to the next step of BIM for WHS management integration. Other changes were made throughout the document, including flagging the matrix earlier, as well as providing a preview of the questions while keeping the actual matrix at the end of the document.

Other additional suggestions were raised. One comment, for example, was to add more case studies, and possibly to focus on critical areas like temporary works and fall from height. By this time, Components 1B and Component 2A had already been stabilised and redrafted: a fourth case study on craneage had already been added to Component 1B; this same case study subsequently laid the groundwork for another Component 2A case study, this time on early contractor involvement. In response to this comment, then, a decision was made to draw out in clearer ways the examples already provided in Component 1A, for example by presenting examples in boxed vignettes. To augment these, additional references were also made to the other components, directing readers to even more real-life examples. Also, rich empirical data was available about craneage, so the research team also decided to focus on this discipline, instead of on temporary works and fall from height, in the development of the last component.

Following the focus group discussion with the Victoria-based participants, a separate interview was conducted with a fourth participant, the NSW work health and safety regulator's representative. Most comments that emerged from this interview were editorial comments, again to sharpen technical details. **Figure 5** shows an example of one participant who was invited to provide direct editorial feedback on Component 1A, specifically on the section on sample Project Information Requirements. The participant raised a few questions to clarify the sample PIR provided. Valuable refinements were also made by this participant in other parts of the document, for example in terms of adding additional descriptors in the Self-Assessment Matrix. Written feedback such as this were combined with other feedback

	timber strip sizes	
PIR 6: A 3D or 4D digital visualisation for each piece of equipment (eg in this example, the crane) shall be developed by the designer progressively throughout design and shared with the Principal Contractor (CDM) and contractor undertaking the lifting operation.	Information on position of perimeter fencing did it include other objects ie buildings or structures within the working zone of the crane Information on position of temporary lifts, the shape and weight of the load? Crane operator information on crane sizes and capacities Did it include the working zone of the crane and people cant work under the crane.	Analysis, planning and simulation

Figure 5.
Sample editorial changes made by participant 4 in component 1A.

obtained in focus groups and the IAG feedback session. In the end, the decision tools evolved to reflect the totality of suggestions captured in the Summary Documents.

Note **Figure 5** has participant changes in purple. In this case, questions raised by the participant in the document were considered alongside comments from other sources. Ultimately, the decision tool drafts were revised to reflect suggestions from a wide range of sources.

The final focus group's aim was to obtain feedback on Component 2B: PIR General and Specific Conditions, which is the most technical element of the Decision Component. Prior to specific discussions on Component 2B itself, participants raised broader issues linked to BIM and WHS management integration. One point of discussion was whether WHS management could be more appropriately understood as a goal of the industry or a by-product of other goals. A second point was whether BIM's ultimate purpose was to achieve better planning and design to reduce risk.

A third point of discussion was the level of specificity to employ in developing this particular set of guidelines. For example, the research team had decided to focus on Significant and Complex Lifts and had developed a number of Specific Conditions in relation to this area. Initially, the Specific Conditions were framed in prescriptive language, with "complex" lifts being defined as loads beyond a certain level of tonnage. This decision was patterned after documents provided by the UK regulator. Through discussions with the Focus Group 4 participants, the research team reached the consensus that "complexity" was more fully understood as an outcome of a range of factors: tonnage, nature of access to the site, dimensions of the load, position of the crane relative to load and capacity of the ground to support heavy cranes. As such, the research team and the focus group participants agreed that it would be counterproductive to define complexity in a strict and narrow sense. The recommendation that emerged was to provide high level examples of complex lifts without prescribing an exact definition, thus allowing the supply chain to decide which lifts were to be classified as complex. Editorial comments, along with content-related comments about the introduction and the General Conditions section, were straightforward. More time was spent on suggestions for Specific Conditions.

One set of discussions involved broadening the list of Specific Conditions to include activities preparatory to lifting, for example the identification of latent conditions and the transportation of elements to the site. Participants also discussed extensively about the type of "lift" that should be subjected to the specific conditions, with both participants pointing out that only lifts with a certain level of complexity would have to be planned using BIM. Complexity could be driven by factors such as size of the structural element, weight, shape, transport considerations or access to site. As one participant pointed out, other projects involving "typical" lifts would not have to be supported by BIM. A project in Barangaroo South in Sydney was raised to illustrate the point: it involved nine tower cranes on the project at its peak and recorded 254,000 crane load lifts in a two-year period, but since these were non-complex lifts, BIM was not required. Because of this, participants also suggested that highly prescriptive terms be avoided, because defining specifics about tonnage were seen to be distracting. When tonnage specifics had to be retained, they had to be changed to ensure that guidance was relevant to the Australian industry. Following the focus group, participants provided further editorial written comments along these lines. One outcome was that the Specific Conditions section was renamed "Complex and Significant Lifts".

When the focus group series was completed, the research team implemented revisions. The revised components were finalised and then uploaded to various websites [25, 26].

4. Discussion and case study impact analysis

Having designed the components and iteratively refined these through three rounds of feedback, the research team then sought to assess the potential impact of each of the components. These impacts, which can also be understood as the benefits of the Decision Framework and as key contributions of this entire study, are analysed in this Discussion section. Specifically, this Discussion section presents an impact analysis of the Decision Framework based on two case studies, referred to here as Case Study 1 and Case Study 2. Assessment of each case study was conducted in three stages: (1) the guidelines and ideals of each component were distilled into 26 questions; (2) the questions were used to analyse project requirements as captured in OIR, PIR and EIR documents in a real case; (3) an assessment was made about the extent to which a specific component would have made a positive difference in the quality of the information requirements. Components have been ordered so that the flow of questions broadly reflect sequential project activities. **Table 10** case study protocol for validating impact of decision framework is a selection of the 26 questions used for the analysis.

Case Study 1 is one of the projects previously analysed in Technical Report 2 and involved the design and construction of an underground metropolitan train station, pedestrian station access and two retail and residential towers. As discussed in Technical Report 2 [15], the project was awarded to a global financial group through an unsolicited proposal; the global financial group then appointed a multinational as the design-and-construct partner for the project. There were significant interactions between the multinational and the government client responsible for transport infrastructure (Government Client [Transport]) because of the station component. It should also be noted that the Government Client (Transport) has over the last decade pursued a digital engineering strategy in its infrastructure projects.

Component	Question for evaluating OIR, PIR, EIR
Component 1A	a. To what extent did the client define a general position on BIM for WHS management, to include in OIR, AIR and PIR, or in equivalent strategic policy documents? b. To what extent did the Client seek feedback from stakeholders to identify risks, opportunities and priorities for the use of a BIM environment to address WHS issues?
Component 2A	a. To what extent did the procurement model for this project support collaborative use of BIM for WHS? b. How were the tender evaluation criteria and weightings selected?
Component 2B	a. How did the Client balance between general and specific requirements, leading to specific priority uses of BIM to address WHS management issues? b. To what extent were general conditions on BIM for WHS defined as part of project information requirements?
Component 1A	a. To what extent was BIM used for WHS in scenario planning, requirement briefing, risk assessment, education and communication, monitoring, reporting?

Table 10.
Case study protocol for validating impact of decision framework.

Case Study 2 involves the design and construction of a 16-kilometre dual carriage motorway in Sydney's west, with a total project cost of $2039 million. Construction is expected to commence in 2022 and the motorway is expected to open by the end of 2025. The motorway is expected to serve as the major access route to the new Western Sydney airport. The project was split into different packages; one project document reported two of these were under a Construct Only procurement model while a third was Design and Construct. Like Case Study 1, the client was Government Client (Transport). As mentioned in the Methodology section, Case Study 2 was not one of the three projects analysed in Phase 2 of the project. To obtain data, the research team examined four project documents and interviewed two project team members: The Technical Director/Digital Engineering Lead and the Digital Engineering Manager.

To illustrate how the desktop review was conducted in response to each of the questions, the following example is provided in relation to Case Study 1, Component 2A. Readers of Component 2A will note the repeated references to the need to require the use of model for WHS management. This was highlighted in

- the case study that explained "an implicit and explicit integration of BIM and WHS management, with 3D models being used in sophisticated ways to plan crane positions, identify dangerous work areas and communicate the sequence of activities"

- the Checklist of Actions to Consider During Tendering (Item 7, Point 2): "Identify examples of how BIM can be used for WHS management".

Thus, Component 2A ideals lead to the framing of the question "How, if at all, were supply chain participants required by the Client to use a model for WHS management (contractual documents, etc.)?" This question can thus be found as Question 2A(g) of the protocol. A total of 26 questions were developed using a similar process.

Responding to these 26 questions required the team to analyse an OIR, PIE or EIR document, or a combination of documents. To address Question 2A(g), the research team analysed one PIR document issued by Government Client [Transport] called the Digital Engineering Standard (Part 1). Analysis took place in three steps.

First, examples of relevant statements in case Study 1 documents that appeared to address the question were identified (Column 1). In this illustration, the PIR document under review had one table entitled "information on possible uses and benefits of Digital Engineering." Specifically, this table showed that BIM models *can be* (not "should be") used for constructability: "BIM models can be used for safety planning, e.g., to analyse construction site layout, including understanding the impacts and interactions with the surrounding areas". These statements, which are sample "evidence", were included in Column 1.

Second, an assessment was developed about the extent to which this example of information requirements demonstrated meaningful and mandatory BIM-WHS management integration. Based on the evidence available in the documents provided, the research team concluded that the client *did not clearly mandate* using of a model for WHS management. Note that this was based on documents that had been made available to the research team; other project documents may exist that may suggest otherwise. But based on the data that was available at the time of analysis, the research team concluded:

- There was, based on Component 2A guidelines, at least one area of strength: The PIR did say that "Use of models is suggested for a range of goals: constructability, quality, optimising construction sequencing", which was favourable.

There were, based on Component 2A guidelines, at least three areas of improvement:

1. Use of the term "Constructability" laid the groundwork for WHS but stopped short of making WHS an explicit goal. "Constructible" does not necessarily mean "safe".

2. Statements were made saying digital information could be used for "analysing a construction site layout". However, statements did not explicit state that this analysis was for the sake of work health and safety purposes, such as identifying risk areas, overworks or underworks.

3. It was also noted that the table referred to "possible uses" of BIM, but not "mandatory" use of models.

Having noted potential areas of improvement, the research team then theorised on what difference Component 2A would have made, had it been available when information requirements were developed. Column 3 thus seeks to address the question: what could have been different in the information requirements if the component had been available as guidance? In response to Question 2A(g), then, the team's response was:

- Component 2A would have *required* clients to facilitate the creation, use and management of the model.

- Component 2A would have driven the client to formalise and mandate the supply chain participants to use digital models in specific ways. For example, the checklist on tendering proficiency would have required clients to (1) identify WHS risks using Best Practice BIM and (2) identify and prioritise key WHS management issues throughout the supply chain (that must be resolved through a BIM environment).

- Component 1B would have provided six potential areas (Knowledge Domains) for the use of BIM models.

- In addition, Component 1A would have also provided additional guidance, for example that a viable starting point for using BIM models for WHS management is using 4D for work sequencing and methodology.

Findings from this impact analysis for both Case Study 1 and Case Study 2 are summarised below. It should be noted that the research team based its analysis on project documents provided (and, in Case study 2, additional interviews). It is possible that some issues identified by the research team as a "gap", "challenge" or "area of opportunity" may in fact be an item discussed and even resolved in a document we did not have access to. In response, we re-emphasise our main purpose: to walk readers through our process of impact analysis that shows possible contributions of the Decision Framework. To support this, we developed two detailed tables with each question answered for each component across three areas: 1) documents 2) assessment and 3) potential contribution/impact. This indicates the transparency of our movement from data to interpretation to assessment. The Technical Report 3 provides the full assessment tables and the following is a summary of the impact analysis.

4.1 Case study 1 impact analysis

Briefly, the findings suggest that the Decision Framework has the potential to strengthen the information requirements developed in the following ways:

- A strong and novel emphasis would have been placed on BIM/WHS management integration. High-level Organisational Information Requirement documents indicated an emphasis on safety, but not on BIM. In contrast, the Project Information Requirement and Exchange Information Requirement indicated high support for the use of digital data, but this emphasis was not explicitly linked to safety goals. Multiple references were made to using digital data to make the construction processes "better", specifically in terms of improved constructability and more efficient sequencing, but how these would translate into better safety performance outcomes, for example a reduced critical incident frequency rate, was not made explicit. This could be reflective of a common yet debatable stance in the industry: that safety is a by-product of better design and construction, rather than a goal in itself. Another example is that project data schemas and project data building blocks were specified to support objectives related to time and cost savings, but not safety objectives.

- A broader swathe of strategic goals, including identifying, recruiting, developing and monitoring supply chain partners, redesigning tendering criteria to capture work health and safety goals and creating environments where even subcontractors could catalyse BIM for WHS innovations, would have been taken into consideration in a much more explicit manner if the Decision Framework was used.

- The use of digital models would have been mandated in systematically identified domains from the outset of the project. The use of 4D as a starting point would have been emphasised. Instead, information requirements that were analysed were found to frame model use as "nice to have" even in critical activities, with specifications saying that "4D models (3D models + time) *may* be used" or that "3D models *can* be used for safety planning".

4.2 Case study 2 impact analysis

Before proceeding directly into findings from Case Study 2, it is important to provide context. Readers will note that at the time of the study, Case Study 2 involved an ongoing project where construction was yet to commence. In contrast, Case Study 1 took place a few years earlier. The development of information requirements and activities such as tendering took place even earlier.

The intervening years between projects is significant because, from a digital engineering perspective, many changes took place within the client's information regime. Some changes came from outside the client organisation. From an international standards perspective, ISO 19650 had not yet been published during Case Study 1, so PIRs had not yet moved to the foreground, and the project team was still expected to submit Employer's Information Requirements. Versions of the EIR template also evolved quite rapidly; one contractor interviewee from Case Study 1 recalled using a "Version 2" template to guide their submission, but at the time of the interview, the research team was provided with an updated "Version 5". EIRs were not provided for Case Study 2, presumably because they are no longer relevant.

Many changes in the overarching information regime, however, were also driven internally by the client itself, with a number of them emerging from lessons that were learned from past projects then transferred to later projects. It is plausible, then, that findings from early projects like Case Study 1 informed Case Study 2. Of course, care must be taken to note the limits of comparisons. The case studies cannot be directly compared in many ways due in no small part to differences in asset types. It should also be noted that an exhaustive comparison of our two cases is beyond the scope of this study. Nevertheless, juxtaposing the two in a considered manner allows some key markers of progress to be discerned. This findings subsection for Case Study 2 is thus organised in two parts, showing (a) progress since Case Study 1, as well as (b) further opportunities for progress identified through the use of the Decision Framework.

4.2.1 Progress since case study 1

Based on high-level comparisons to Case Study 1, key areas of progress in DE-enabled WHS management discerned in Case Study 2 are listed below.

- WHS project objectives are now explicitly included in the PIR (the quality specifications document), in the form of "Improved safety in construction (improved HSiD)", with HSiD referring to "Health and Safety in Design". Decision-making in support of this objective is buttressed by information management classifications and codes and the use of a software called BIM Track, a point that is raised in the next bullet. In contrast, Case Study 1 documents (contractor DEXP and EIR) showed nominal references to Health and Safety, with the contractor DEXP showing general statements ("Digital processes will be required to be used for Safety in Design Reviews (SiDRs) throughout the Design Phase") or non-prescriptive commitments ("4D models [3D + time] may also be created where there are critical risk installations").

- "Safety" is now a type of metadata identifier, a code, a business discipline classification and a BIM Track type issue. No references to this type of tagging were found in Case Study 1 documents. All of these information conventions point to this supply chain's relatively stronger capacity for more accurate and efficient identification of safety issues as well as the potential for more systematic accumulation, management and analysis of safety-related data.

- Greater intentionality can be discerned in terms of how the client balances between flexibility and specificity in requiring the use of digital engineering to support WHS objectives. For example, the Government Client (Transport) includes a list of model uses in its PIR, including Health and Safety in Design analysis. However, instead of requiring all of these model uses, the PIR instead notes that the "DEXP shall identify which of the following uses will be supported", thus making the conscious choice to allow tenderers to make certain decisions themselves. This is consistent with the recommendations of Component 2A, which notes the benefits of defining key requirements while leaving space for innovative responses.

- Requirements for the use of specific BIM capabilities like 4D are now more directed and informed. One of the findings that emerged during an earlier contractor interview for Case Study 1 was that Government Client (Transport) specified too many requirements without knowing how it would use these. In Case Study 2, requirements were much more focused, possibly because the client had drawn on past learnings. The use of 4D was scoped, with documents showing that it was required in situations where it was deemed "appropriate, [capable of] improving visual communication and optimisation of construction staging and sequencing." Follow up interviews confirmed this, with interview participants noting that 4D was used in design reviews of selected locations only, like complex interchanges.

- A richer range of collaborative meetings appears to have been planned. In Case Study 1, three types of meetings were defined in the DEXP, and these meetings included BIM managers, design managers and the building owner/developer; documents provided made no mention of WHS professionals or those with experience and knowledge in identifying safety risks participating. In contrast, Case Study 2 DEXP shows six different types of meetings, including safety in design meetings for each subdiscipline led by the main contractor's health and safety lead manager. Interview data also confirmed that WHS professionals were regular participants in these collaborative meetings, including early design meetings where the use of integrated models prompted the identification of WHS issues.

- One interesting point that emerged during interviews was the regular inclusion of asset managers in monthly meetings. One interviewee also noted that the federated model of the final design must now be approved by asset managers before construction commences. These practices suggest that Government Client (Transport) is now taking a more considered approach that looks beyond using DE for design and construction to asset management. This approach could lay the groundwork for improved WHS during the use phase of the asset.

4.2.2 Further opportunities identified in case study 2

While progress has been made in key areas, opportunities for improved DE-enabled WHS management have nevertheless been identified using the guidelines and ideals distilled in the Decision Framework. Some examples follow:

- While project-level goals involving DE-enabled WHS management were clear, higher-level goals, ideally expressed in an OIR, did not appear, at least not in ways that integrated DE and WHS. Documents on Government Client (Transport)'s long-term strategy referred to safety, but not to DE-enabled WHS. Component 1A of the Decision Framework recommends foregrounding "integrated" BIM/DE for WHS management consistently across all tiers of goals, for example, in project goals as well as in organisational objectives that transcend project goals. Those who familiarise themselves with the recommendations of Component 1A would thus be sensitised to the fact that in this case study, BIM/WHS integration is emerging at the project level but remains inconsistent at the higher-level vision- and organisational levels. This lack of integration could be reflective of a persistent dilemma in the industry discussed in Focus Group Interview 4, where participants noted that WHS could be a goal in itself or a means to other goals.

- Supply chain capability requirements for the project were clearly not defined during tendering. The PIR simply notes that it is a requirement for tenderers to use DE models, which "must be supported by the Contractor's DE." The phrasing appears to simply assume that the supply chain should have the capability to use DE for a range of goals. Thus, the client requires the "ability to use models for a range of purposes" as part of the capacities of the supply chain that are required for this project but does not specify required knowledge, skills and competencies. Follow up interviews confirm this as an area for improvement. The only type of supply chain capability assessment that was carried out involved a request for the CV of the tenderer's DE person, a practice that has now proven to be inadequate. Guidelines in Component 1A suggest that clients must go beyond assuming supply chain capability; instead, clients must explicate skill and competency expectations, including those related to BIM for WHS. Clients are, in fact, encouraged to participate in more proactive ways, for example, in building industry-wide capability and even introducing new roles.

- While WHS professionals have been found to be involved in early project stages, for example, in early design reviews, there is still no clear evidence that they participate in tender evaluation. Component 2A recommends that tender evaluation panels should be diversified to include professionals capable of assessing a supply chain's WHS methods, capabilities and performance. On the related matter of tender assessment criteria, there is no evidence that allowed the research team to assess the weight of DE-enabled WHS management capabilities in tender assessment.

- While the client's overall strategy of balancing between flexibility and prescriptiveness is sound, judging where to "draw the line" between the two is a capability that is sharpened over time, as lessons learned accumulate. Hence there is still room for the Government Client (Transport) to strengthen its capacity to define where this line should rest. For example, for this project, the Government Client (Transport) set minimum requirements for clash detection, use of 4D to identify constructability issues, information exchanges, model federation, and models required. This practice of "minimum" standard-setting may be in place because the Government Client (Transport) is still "climbing the learning curve" in many areas; as one interviewee noted, "we're not experts on WHS", thus it becomes strategic to set a baseline and allow more innovative tenderers to propose their own novel solutions. However, in some cases the choice of using minimum standards has led to rather generic contractor responses. In Case Study 2, the main contractor had very broad descriptions of how they planned to use 3D and 4D: for example, "3D visualization will be provided in accordance with the detail design deliverables in the form of digital renders and animations (and) will be undertaken by a member of the modelling team" and "4D construction sequencing visualization will be provided in accordance with the detail design deliverables in the form of digital animations." Component 2A recommendations could help address this challenge of lack of specificity in responses. If the Government Client's tendering criteria had been broadened to include BIM for WHS components, as recommended in Component 2A, then tender evaluators would have been better sensitised to note that the main contractor's broad response did not fully capture the client's target to use 4D for "issues and assets that are complex or otherwise warrant close attention to prove constructability which is safe, efficient and practical".

- While the PIR specifies for the submission of a pre-contract DEXP, assessment of this submission has been described as being something of "a light touch". Documentary evidence suggests that the Government Client (Transport) expects the pre-contract DEXP to include many things, including explanations about the structured sharing and use of digital data for things like clash detection. These clearly lay the groundwork for the use of DE to support WHS. The pre-DEXP is also expected to include explanations on the "(f)requency of collaboration and information exchange noting that [the client] requires the ability to progressively view and discuss progress on models in addition to the specified review milestones" as well as "(d)etails of model workshops and other collaborative working practices". While these details were included in submissions, follow up interviews indicate that assessment may not have been done rigorously. One interviewee noted that "smart" tenderers could simply "cut and paste" what is in the client's digital specifications into their tender responses. When tender responses are evaluated, and while "objective" assessment criteria are provided, some aspects of tender evaluation have been described as "subjective" and limited to seeing "whether an item was (simply) mentioned or not".

- Digital data management is still suboptimised despite the presence of "Safety" as a metadata identifier allowing the tagging and collection of safety data. Data is collected in spreadsheets. Government Client (Transport) has not yet begun the practice of incorporating work health and safety data into the model. Component 2A highlights the need for establishing the currency and accuracy of WHS data, and while there is emerging evidence of the ability to build a repository of WHS information through the management of issues, the full potential of digitally-supported WHS decision-making has not yet been attained. Still, the evidence shows promising beginnings. The current approach of tagging and labelling data is consistent with the recommendation in Component 2B for the establishment of a structured, standardised data schema that enables the design and related WHS information to be accessed, filtered and used by other participants in planning, managing and controlling WHS risks.

Similar to Case Study 1 detailed impact analysis was completed.

5. Conclusion

The challenge to all participants in the construction industry is to find ways to reduce WHS incidents and injuries to support a safe workplace. In Australia, construction industry participants and clients have significant legislation and accountabilities.

BIM, as a process, can facilitate the early identification of potential WHS issues and the application of preventative strategies using automated approaches. The overarching benefit of utilising BIM for WHS management is to ensure that all work health and safety data and information is managed within the one data repository. Apart from this strength of managing and assessing information from all parties involved in WHS; the strengths of using BIM for WHS management manifest in additional contributions such as the embedded processes of visualisation, simulation, analytics, evaluation and monitoring. BIM also supports communication, education and training during the design, construction, operational phases and post construction phases. Documenting WHS requirements in plans and monitoring compliance with WHS regulations within a BIM-environment can result in safer working conditions as WHS is identified as an integral part of the construction process, is made more visible to stakeholders and can be appropriately managed at the various stages of construction.

Internationally we lack research that evaluates WHS management in a BIM environment. The aim of our four-phase study was to explore practices reflecting BIM-enabled WHS management. Such practices have not been underpinned by comprehensive concerted studies underpinning these practices. A key outcome of this study has been to analyse BIM for WHS management in practice and through that rigorous analysis develop sound recommendations for industry. Further to that, the recommendation to apply the research findings to the development of useful and practical guidelines specifically targeting BIM and WHS management was enacted and described. This chapter's focus has been to document the systematic process of applying findings from previous phases to develop decision tools. This represents a rigorous process of knowledge transfer between academia, industry and government.

Our findings indicate that the management of WHS is likely to be most effective when considered holistically as part of the project lifecycle, with clear directional client leadership in the early stages of planning a project and then through the project tender and evaluation stage and thus enacted in a BIM-enabled environment by the project leadership team of contractor, consultants and key major subcontractors.

BIM-enabled WHS management is rarely considered in construction tender requirements and the evaluation of bidders' proposals in an explicit manner, despite the scope for BIM to provide a powerful approach to the management of WHS and despite the crucial desire to improve work safety environments. Such clear expectations allow WHS requirements to be embedded early in the design and construction process and facilitates increased clarity for the management of WHS across the project lifecycle.

The UK, Singapore, Hong Kong and Finland public and private sector clients appear to be further advanced in the adoption of BIM and the overarching benefits use for WHS management. They provided valuable research sites that could enrich information on the approaches, tools and outcomes highlighted by the theoretical, prototype and evaluative studies in this area. Sufficient examples of Australian private and public sector BIM-enabled construction projects now exist to allow evaluation of BIM, from the tendering stage through to post construction, and the potential creation of management and information systems for monitoring WHS. There also appears to be sufficient experience among industry and government stakeholders for major infrastructure projects to develop models for the adoption of BIM-enabled WHS management systems. In Australia, the development of the Victorian Digital Asset Strategy and the NSW Information Management Development Framework are two examples showing that agencies are maturing and are becoming increasingly capable of creating frameworks that support BIM and WHS management integration. There are also other documents like the 10 Point Commitment to the Construction Sector [21] that focus on creating collaborative environments that support holistic BIM adoption in ways that support WHS management. There are many other such strategic documents that are important and which we discussed in detail in Phase 1 Technical Report. That said, there is also a need for tools and guidelines that bridge these high-level ideals with operational targets. The Decision Framework developed for this phase of the research is an attempt to bridge this gap and a first step at translating lofty strategic goals into concrete actions.

Clearly, the brave new world of BIM holds potential for supporting the fulfilment of WHS obligations and it is timely to harness the potential of BIM to ensure safer construction sites. The suite of guidelines is the beginning of a new approach to improving adoption of BIM in our industry and the next step is to develop more detailed guidelines and standards at the project level to support the strategic approach presented in this study. It is time we did better.

Acknowledgements

- Centre for Work Health and Safety and in particular Angelica Varhammar; Ron Keelty, SafeWork NSW, Western Sydney University Project engagement team participants, Industry Advisory Group members, interviewee participants who offered their experiences, time and ideas

- Prof. Ning Gu (University of South Australia) and Ms. Claudelle Taylor (CIMIC Group Limited) who reviewed the final report

- Gordon Crick UK HSE for providing documents and thoughtful insight into the UK progress towards BIM for WHS integration

Author details

Kerry London* and Zelinna Pablo
Torrens University Australia, Sydney, Australia

*Address all correspondence to: kerry.london@torrens.edu.au

References

[1] Garner-Purkis Z. Deaths in Construction Rise 27% Year on Year. UK: Construction News; 2018. Available from: https://www.constructionnews.co.uk/best-practice/health-and-safety/deathsin-construction-rise-27-year-on-year/10032802.article

[2] Bureau of Labor Statistics (BLS). National Census of Fatal Occupational Injuries. US: Bureau of Labor Statistics (BLS); 2019. Available from: https://www.bls.gov/news.release/pdf/cfoi.pdf

[3] International Labour Organisation. World Statistics. US: International Labour Organisation; 2020. Available from: https://www.ilo.org/moscow/areas-of-work/occupational-safety-and-health/WCMS_249278/lang--en/index.htm

[4] Safe Work Australia. Australian Work Health and Safety Strategy 2012-2022. NSW Australia: Safe Work Australia; 2020. Available from: https://www.safeworkaustralia.gov.au/about-us/australian-work-health-and-safety-strategy-2012-2022#

[5] London K, Pablo Z, Feng Y, Rahnamayiezekavat PR, Varhammar A, Zhang P. Health and Safety Management Using Building Information Modelling: Phase Three Report. NSW Australia: NSW Government; 2021. p. 130. Available from: https://www.centreforwhs.nsw.gov.au/__data/assets/pdf_file/0005/1036895/Centre-for-WHS-BIM-Project-Phase-3-Report.pdf [Accessed: October 11, 2024]

[6] London K, Pablo Z, Kaur. VA, Feng Y, Zhang P. Health and Safety Management Using Building Information Modelling: Phase One Report. NSW Australia: Government of New South Wales; 2020. Available from: https://apo.org.au/node/309518

[7] Akram R, Thaheem MJ, Nasir AR, Ali TH, Khan S. Exploring the role of building information modeling in construction safety through science mapping. Safety Science. 2019;**120**:456-470

[8] Azhar S. Role of visualization technologies in safety planning and management at construction job sites. Procedia Engineering. 2017;**171**:215-226

[9] Lingard H, Oswald DLT. Embedding occupational health and safety in the procurement and management of infrastructure projects: Institutional logics at play in the context of new public management. Construction Management and Economics. 2019;**37**(10):567-583. DOI: 10.1080/01446193.2018.1551617

[10] London K, Singh V, Taylor C, Gu N. Towards the development of a project decision support framework for adoption of building information modelling. In: Underwood J, Isikdag U, editors. Handbook of Research on Building Information Modeling and Construction Informatics: Concepts and Technologies. Hershey, PA: IGI Global; 2010. pp. 270-301

[11] Martínez-Aires MD, López-Alonso M, Martínez-Rojas M. Building information modeling and safety management: A systematic review. Safety Science. 2018;**101**:11-18

[12] Tan T, Chen K, Xue F, Lu W. Barriers to building information modeling (BIM) implementation in China's prefabricated construction:

An interpretive structural modeling (ISM) approach. Journal of Cleaner Production. 2019;**219**:949-959

[13] del Lopez Puerto C, Clevenger C. Enhancing safety throughout construction using BIM/VDC. In: Presented at ECOBuild 2010 BIM Academic Forum, Washington DC, US. 2010

[14] Hamed G, Banihashemi S, Hon C, Drogemuller R. Building Information Modelling and Workplace Health and Safety. 1st ed. UK: Routledge; 2023. eBook ISBN9781003224853

[15] London K, Pablo Z, Feng Y, Rahnamayiezekavat PR, Varhammar A, Zhang P. Health and Safety Management Using Building Information Modelling: Phase Two Report. Australia: New South Wales. Centre for Work Health and Safety; 2021a. Available from: https://www.centreforwhs.nsw.gov.au/__data/assets/pdf_file/0009/993627/BIM-and-WHS-Phase-2-Report.pdf

[16] London K, Kaur G, Feng Y, Vårhammar A, Wallace G, Zhang P, et al. Construction work health and safety management using building information modelling Ch. 4. In: Underwood J, Shelbourn M, editors. Handbook of Research on Driving Transformational Change in the Digital Built Environment. UK: IGI Global; 2021b. pp. 91-119. DOI: 10.4018/978-1-7998-6600-8.ch004

[17] Gu N, London K. Understanding and facilitating BIM adoption in the AEC industry. Automation in Construction. 2010;**19**(8):988-999

[18] UK BIM Alliance. BIM4H&S Working-Group PIR Template. 2020. Available from: https://airtable.com/shrUnnf1bOJXPJhzU/tblDI61xaEbnpcruL/viwcUqeD9UxOWhUvO?blocks=bipJUW7hnwsqb5lDi

[19] UK BIM Framework. Information Management According to BS EN ISO 19650 Guidance Part D: Developing Information Requirements. 1st ed. UK: BIM Framework; 2020. Available from: https://www.ukbimframework.org/wp-content/uploads/2021/02/Guidance-Part-D_Developing-information-requirements_Edition-2.pdf

[20] UK BIM Framework. Information Management According to BS EN ISO 19650 Guidance Part E Tendering and Appointments. 1st ed. UK: BIM Framework; 2020. Available from: https://www.ukbimframework.org/wp-content/uploads/2021/02/Guidance-Part-E_Tendering-and-appointments_Edition-2.pdf

[21] NSW Government. NSW Government Information Management Framework. Licensed under a Creative Commons Attribution 3.0 Australia License. 2018. Available from: https://data.nsw.gov.au/information-management-framework [Accessed: October 14, 2024]

[22] International Organization for Standardization. ISO 19650-1: 2018 Organization and Digitization of Information about Buildings and Civil Engineering Works, Including Building Information Modelling (BIM) — Information Management Using Building Information Modelling — Part 1: Concepts and Principles., Part 2: Delivery Phase of the Assets. Part 3: Operational Phase of the Assets. Part 4: Information Exchange. Part 5: Security-Minded Approach to Information Management. Geneva: ISO; 2018

[23] Guidelines for the Integration of Work Health and Safety and Building Information Modelling Management. Available from: https://www.centreforwhs.nsw.gov.au/tools/guidelines-for-the-integration-of-whs-management-in-bim [Accessed: October 14, 2024]

[24] London K, Pablo Z, Feng Y, Rahnamayiezekavat PR, Varhammar A, Zhang P. Health and Safety Management Using Building Information Modelling: Phase Three Report. NSW Australia: NSW Government; 2022. Available from https://www.centreforwhs.nsw.gov.au/__data/assets/pdf_file/0005/1036895/Centre-for-WHS-BIM-Project-Phase-3-Report.pdf

[25] Torrens University Featured Research Computer Modelling to Improve Safety on Sites. Available from: https://www.torrens.edu.au/research/featured-research/computer-modelling-to-improve-safety-on-construction-sites

[26] Building Information Modelling for Work Health and Safety Management Research Project Webpage. Available from: https://www.centreforwhs.nsw.gov.au/research/building-information-modelling-for-whs-management [Accessed: October 14, 2024]

Chapter 3

Synergies between Integrated Project Delivery and Big Open BIM

Daniel Krause, Claudia Szargan and Maria Zuluaga

Abstract

The "Neues Werk Cottbus" project is Germany's first major initiative to apply the new Integrated Project Delivery (IPD) contract model in combination with BIM methodology. The project alliance, consisting of the client, designers, and construction companies, provides optimal conditions for implementing an innovative, interdisciplinary Big Open BIM approach from the outset. The main objective was to develop a versatile, project-wide Project Information System (PIS) compatible with various software systems. This PIS supports over 20 BIM use cases, ranging from designing the maintenance facility and assembly planning to construction execution and building operation. This comprehensive approach ensures consistent generation, use, and updating project data throughout the entire building lifecycle without redundancies. Multidisciplinary collaboration has proven extremely valuable during design and construction. Notable benefits include greater transparency in model-based planning and meetings, measurable process improvements through detailed 4D scheduling simulations, more precise target price and cost management using model data, and significant efficiency gains through cloud-based, workflow-oriented collaboration. By implementing open BIM process standards and data formats, the project demonstrates how IPD and Open BIM methodologies reinforce each other, enabling seamless interdisciplinary collaboration. This project sets a benchmark for digital construction and lifecycle-oriented project management in Germany.

Keywords: building information modeling, integrated project delivery, openBIM, infrastructure construction, project alliance, common data environment

1. Introduction

The project "Neues Werk Cottbus," which involves constructing a modern railway maintenance facility for Deutsche Bahn's growing long-distance train fleet, is the first major project in Germany to apply the new Integrated Project Delivery (IPD) contract model while simultaneously using BIM methodology. The project alliance, consisting of the client, designers, and construction companies, provides the ideal conditions to develop and implement an innovative, holistic, and interdisciplinary Big Open BIM approach from the very beginning.

This multi-disciplinary collaboration has already been proven extremely valuable in the design and construction phases. Examples of added value include higher transparency in model-based planning and construction meetings, process improvements through detailed 4D simulations in scheduling and work preparation, better target price determination and cost management using model data, and significant efficiency gains through project-wide collaboration in digital cloud applications. All this would not have been possible without the implementation of an integrated system approach using openBIM process standards and data formats.

The "Neues Werk Cottbus" project is a significant step forward in BIM implementation for all participating companies and an outstanding example of successful collaboration with innovative technologies. It demonstrates that through a holistic Open BIM approach and the use of open data standards in combination with partnership-based contract models, large-scale construction projects can be executed with greater efficiency, precision, and teamwork.

2. Project description

"Neues Werk Cottbus" is not just another construction project; it represents a critical initiative involving the development of two maintenance halls for high-speed electric trains, which are essential for Deutsche Bahn's operations. As Deutsche Bahn's long-distance fleet expands to 450 ICE trains by the end of the decade, the demand for efficient maintenance will present a significant strategic challenge. The new ICE trains must undergo their first major service six years after entering operation to continue running reliably. To facilitate this, new maintenance tracks must be available starting in 2024. Thus, this project is a cornerstone in ensuring that these trains are rapidly returned to service while also playing a vital role in supporting the mobility transition and advancing climate protection in the transportation sector.

The project entails the construction of two major maintenance facilities shown in **Figure 1**. The first maintenance hall, which measures 445 meters in length and 33 meters in width, was completed and inaugurated at the beginning of 2024, marking its construction a record-breaking achievement. The second maintenance hall, currently under construction, is set to be even larger, with a length of 526 meters and a width of 200 meters. In addition to the main hall, six additional buildings are planned as part of the overall facility, with the entire complex scheduled for completion by 2026.

What is particularly noteworthy about this project is the exceptional speed at which it is being executed. The first double-track maintenance hall was completed and became operational in less than two years—from groundbreaking to commissioning in early 2024. This rapid timeline was driven by the urgent need for maintenance tracks to accommodate the six-year maintenance inspections of the new ICE fleet. Despite widespread supply chain challenges, this impressive timeline was achieved, ensuring that the new ICE trains can be efficiently serviced and quickly returned to operation. This success was made possible by using the Integrated Project Delivery (IPD) partnership contract model in conjunction with an open Building Information Modeling (BIM) approach, which enabled seamless collaboration and efficient coordination across all parties involved, from design to commissioning.

Looking ahead to the construction of the second maintenance hall, the integration of BIM and IPD is expected to further enhance the project's efficiency, thanks to the lessons learned and the standards and processes established during the construction of the first maintenance hall. By leveraging these advanced methodologies, the project team aims to proactively address potential issues and mitigate delays, ensuring the timely delivery of the entire project.

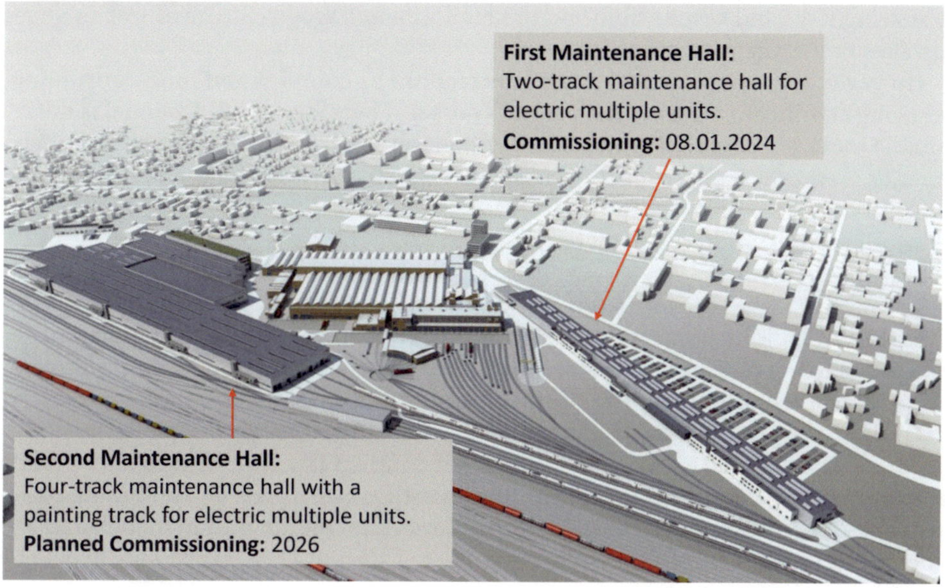

Figure 1.
Overview from maintenance halls.

3. IPD in the project

3.1 Contract model IPD: Definition and principles

The IPD contract model is an innovative approach to executing construction projects that emphasizes cooperation and shared responsibility. All key stakeholders—owners, designers, contractors, and others—collaborate from the project's inception to completion, sharing both risks and rewards. This method fosters transparency, improves efficiency, and reduces waste by aligning the goals of all parties involved.

Remuneration under the IPD contract is based on actual cost reimbursement plus a profit markup. This means that all incurred costs are reimbursed, and a predetermined profit markup is also provided. This structure incentivizes everyone involved to work efficiently and minimize costs.

The basis for remuneration is the target price, which is jointly determined by all partners based on collaboratively developed plans. The expertise of both the planners and the construction companies is considered. This target price serves as a reference point for cost control. Additionally, a risk budget is defined to cover unforeseen costs. This budget is managed collectively by all project participants to minimize risks and resolve issues swiftly.

Shared responsibility and the early involvement of all parties promote collaboration. The incentive systems and transparent cost structure help ensure that projects remain within budget and on schedule. Risks are shared, leading to cooperative problem-solving.

Thus, the IPD model provides an ideal foundation for implementing BIM, as its collaborative structure aligns seamlessly with the integrated digital workflows required for BIM. Additionally, IPD supports the use of the Last Planner method, promoting a highly cooperative and efficient approach to planning and scheduling.

By sharing risks and responsibilities, the model encourages a culture of learning from mistakes rather than assigning blame.

However, managing a multi-party contract can be complex and time-consuming, requiring significant effort at the project's outset. Therefore, the IPD model is most suitable for large, complex projects involving multiple trades, and its success largely depends on the level of trust between the involved parties [1–4].

3.2 Implementation in the project

In the early stages of the project, it became evident that completing the two maintenance halls by January 2024 and the end of 2026 would require significant efforts due to the project's extensive scope, the high likelihood of changes, complex technological requirements, and the involvement of numerous stakeholders. For this reason, the project team began implementing the project according to the IPD model in November 2020. With the support of contract law experts who have practical experience with the IPD model and the scientific backing of TU Berlin, the basic principles, organizational structure, scope of services, remuneration model, and a timeline with milestones leading up to the start of the project alliance were developed through several workshops.

The selection of partners followed a multi-stage process, beginning with a competition and the submission of initial bids. Several assessment sessions were held to evaluate the participants' suitability for the project based on predefined criteria. This process culminated in final offers from the bidders and the ultimate award decision.

Simultaneously, a co-working space was established at the Cottbus facility to foster collaboration among project participants. The project alliance officially commenced in January 2024 with a joint workshop for all stakeholders. During the first year, several changes occurred regarding the partners involved. An introduction to the project participants will be provided in the following chapter.

3.3 Key participants

The "Neues Werk Cottbus" project unites eight leading companies, each contributing their expertise toward a shared goal: completing the project within the estimated timeframe while utilizing resources efficiently and effectively. While this chapter focuses on the main companies involved, it is essential to acknowledge the significant role of their subcontractors, who are also crucial to the project's successful development.

The client for this project is DB Fahrzeuginstandhaltung GmbH, a part of the DB Group responsible for maintenance. The planning of the railways and external facilities is managed by FCP GmbH. Arcadis Germany GmbH and Baumert & Peschos GmbH share responsibility for planning the building infrastructure, architecture, and technical building equipment, while LOGSOL GmbH oversees the planning of logistics systems.

The construction of the building structures is carried out by Wayss & Freytag Ingenieurbau AG, and Rhomberg Sersa Rail Group is responsible for the construction of the railways and external facilities. Engie Deutschland GmbH is tasked with the assembly and installation of the technical building equipment.

Although these companies are distinct entities, they operate collaboratively as an alliance, supporting each other's tasks to achieve the best possible outcome for the project (**Figure 2**).

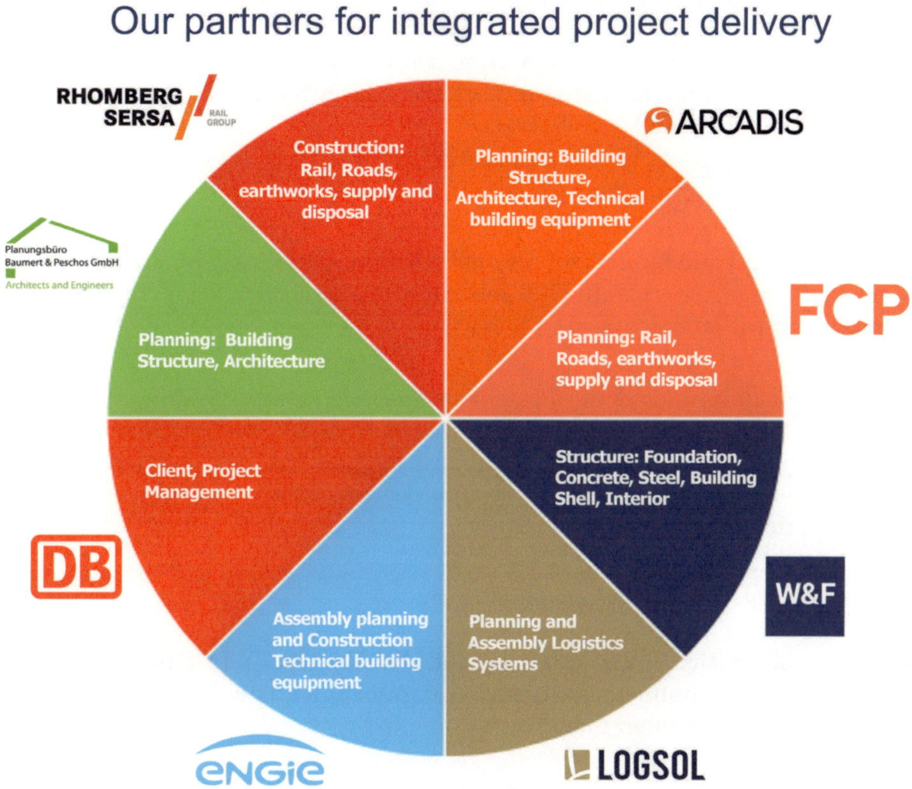

Figure 2.
Key participants overview.

4. BIM in the project

4.1 Goals and objectives

The "Neues Werk Cottbus" project is founded on a well-defined framework of goals and objectives that guide its successful execution. These goals outline the strategic vision, emphasizing the integration of advanced BIM methodologies and innovative collaborative approaches throughout all project phases. The objectives specify measurable actions to achieve this vision, facilitating efficient management, coordination, and execution. By establishing these objectives, the project ensures alignment among stakeholders, optimizes resource utilization, and addresses the complex technical and operational requirements essential for its successful completion.

The objectives determined for the project are as follows.

4.1.1 Develop a versatile cross-project information system (PIS)

Create a robust information system capable of managing diverse BIM use cases, including design, assembly planning, and construction execution, while supporting lifecycle-spanning project management and minimizing data redundancy.

4.1.2 Implement a unified project information structure

Establish a standardized information structure that supports open and integrated management throughout the entire project lifecycle, ensuring consistency and effective collaboration across all stakeholders.

4.1.3 Integrate best software solutions

Carefully select and implement optimal software solutions from project partners to support the lifecycle-spanning PIS and improve project efficiency, aligning technology choices with project needs.

4.1.4 Enhance collaboration through a common data environment (CDE)

Utilize centralized data environments, such as Autodesk Construction Cloud, to streamline collaboration across all project phases, providing stakeholders with up-to-date information in real-time.

4.1.5 Improve visualization and communication

Leverage immersive visualization methods like Resolve VR to enhance stakeholder engagement and communication, ensuring alignment with public authorities and key decision-makers throughout the project.

4.1.6 Develop a cost management workflow

Streamline cost estimation processes by developing efficient workflows using software such as iTWO, allowing for accurate cost management that accounts for the diverse expertise of project participants.

By achieving these goals and objectives, the "Neues Werk Cottbus" project will not only provide a cutting-edge maintenance facility but also set a new standard in the use of BIM and collaborative processes for large-scale infrastructure projects. This comprehensive approach ensures efficient management and coordination across all project stages.

4.2 BIM use cases

The PIS developed for the "Neues Werk Cottbus" project is a comprehensive digital framework that supports over 20 distinct use cases, shown in **Figure 3**, which were created using the buildingSMART use case list as a reference [5, 6]. These use cases represent specific scenarios where BIM methodologies are employed to achieve project objectives. BIM plays a crucial role in this project, facilitating various phases of the building lifecycle—including design, construction, and operation—by specifying the necessary processes, data interactions, and technological integrations. This structured approach ensures efficient management, coordination, and execution across all stages of the project.

The use cases implemented in the "Neues Werk Cottbus" project are organized into five distinct groups, each corresponding to different stages in the project lifecycle. This categorization enhances clarity and allows for focused application of BIM

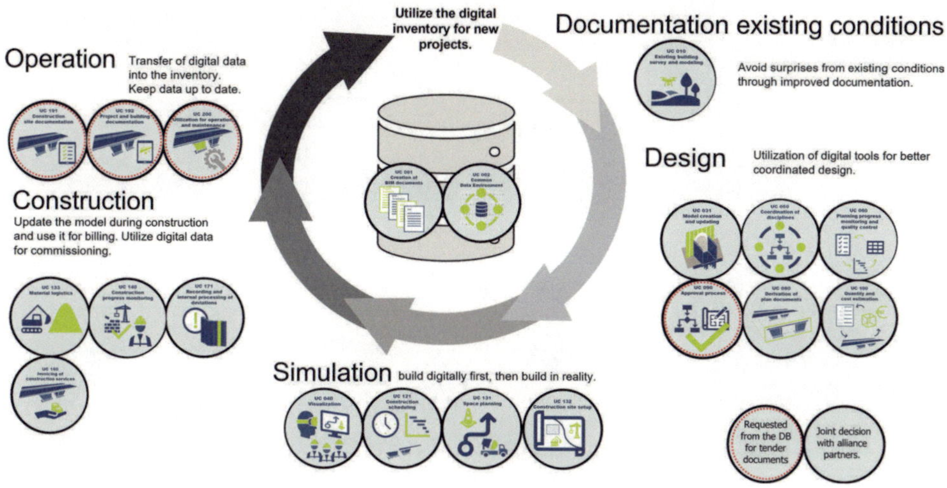

Figure 3.
BIM: Use cases overview.

tools and processes. The use cases are grouped into five categories, corresponding to different stages in the project lifecycle:

4.2.1 Use-case category 1: Documentation of existing conditions

- *Existing building survey and modeling*: Detailed 3D model or point cloud of existing structures integrates surveys, environmental data, and MEP system performance, ensuring accuracy (**Figure 4**).

- *Creation of BIM documents*: BIM documentation, including the BIM Execution Plan (BEP) and Employer's Information Requirements (EIR), standardizes project requirements.

Figure 4.
3D model of the maintenance hall overlaid with a point cloud of the existing.

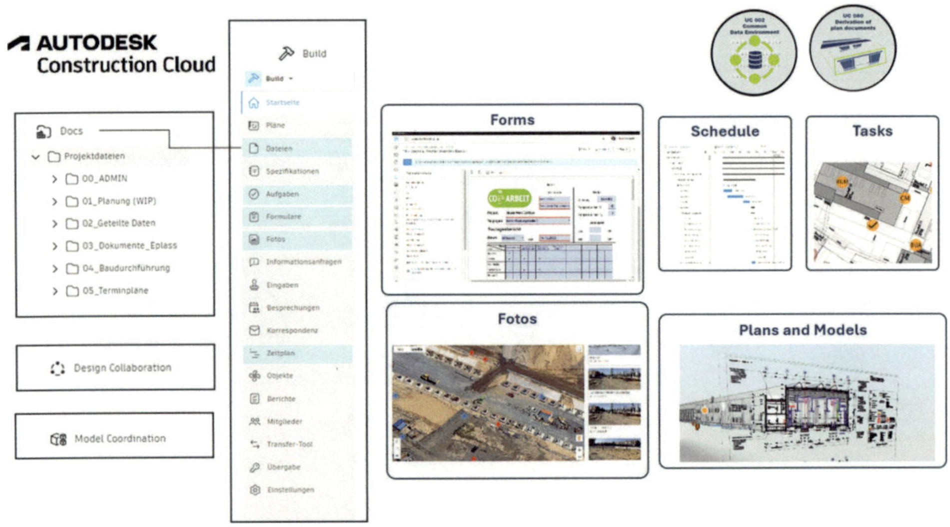

Figure 5.
Autodesk construction cloud interface.

- *Common data environment (CDE)*: A centralized database categorizes project information into In Progress, Shared, and Published, per DIN EN ISO 19650 standards (**Figure 5**).

4.2.2 Use-case category 2: Planning and coordination

- *Model creation and updating*: Discipline-specific 3D models are created, integrated, and updated to form a final coordinated model.

- *Coordination of disciplines*: Ensures alignment and coordination among various project teams using 3D models.

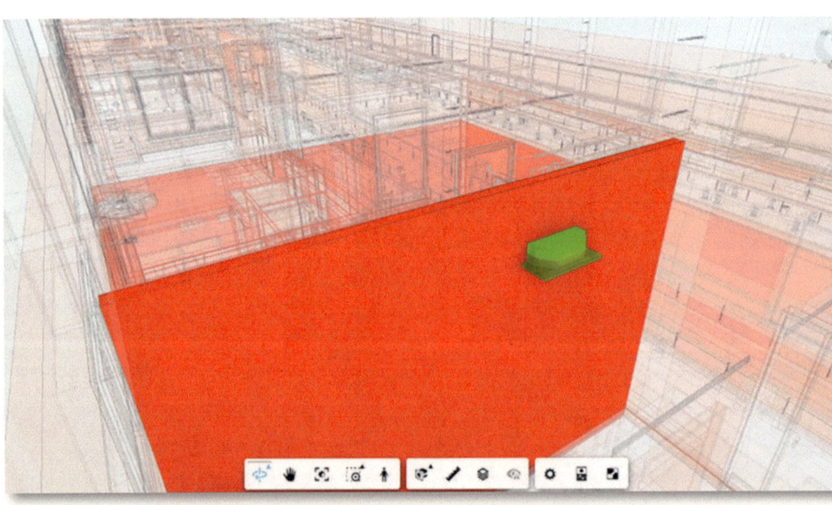

Figure 6.
Clash detection.

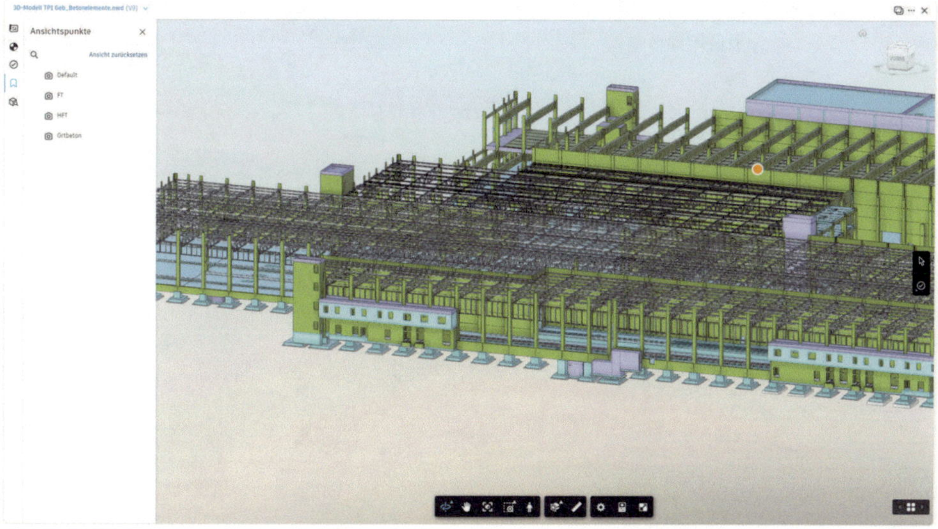

Figure 7.
Color-coded concrete elements sub-model.

- *Clash detection and resolution*: Regular clash detection is performed using integrated discipline-specific 3D models to identify and resolve conflicts early in the design process (**Figure 6**).

- *Derivation of plan documents*: Plan approval processes are streamlined through the CDE, with model-based collaboration supported by BCF interfaces.

4.2.3 Use-case category 3: Simulation and visualization

- *Visualization*: 3D models facilitate project meetings, planning, and public relations by providing detailed visual representations (**Figure 7**).

- *Construction scheduling and space planning*: Simulation links model elements to schedules, optimizing logistics and site planning (**Figure 8**).

- *Construction site setup*: 4D models help plan site infrastructure and traffic management (**Figure 9**).

4.2.4 Use-case category 4: Construction phase

- *Construction progress monitoring*: Schedule-based progress control compares actual progress with plans, informing decision-making (**Figure 10**).

- *Recording and processing deviations*: Clashes are detected and resolved through systematic conflict management in BIM meetings (**Figure 11**).

- *Material logistics*: 4D models support efficient material delivery and handling, minimizing delays.

Figure 8.
Construction site setup model.

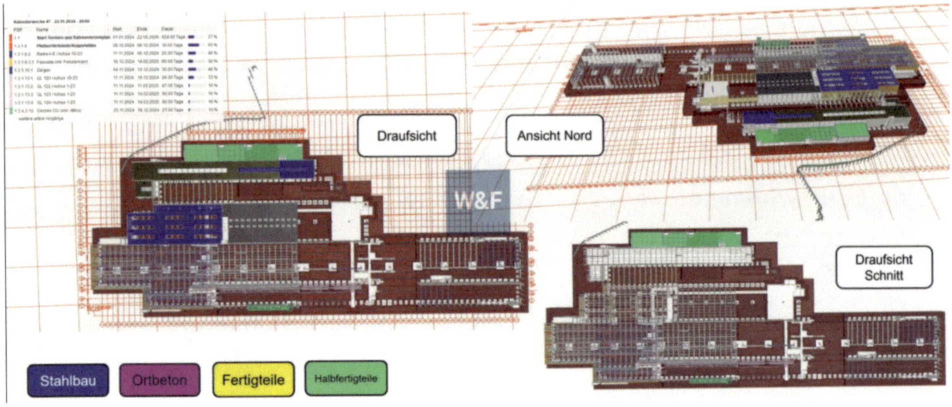

Figure 9.
4D model.

4.2.5 Use-case category 5: Operation and maintenance

- *Construction site and project documentation*: Comprehensive documentation through advanced project management software, including GPS-referenced photos and daily reports.

- *Utilization for operation and maintenance*: The as-built BIM model continues to support facility management and maintenance beyond construction.

By organizing over 20 distinct use cases into focused categories, the project leverages BIM to enhance coordination, streamline processes, and improve decision-making from initial documentation to operation and maintenance. This structured

Figure 10.
Construction progress monitoring.

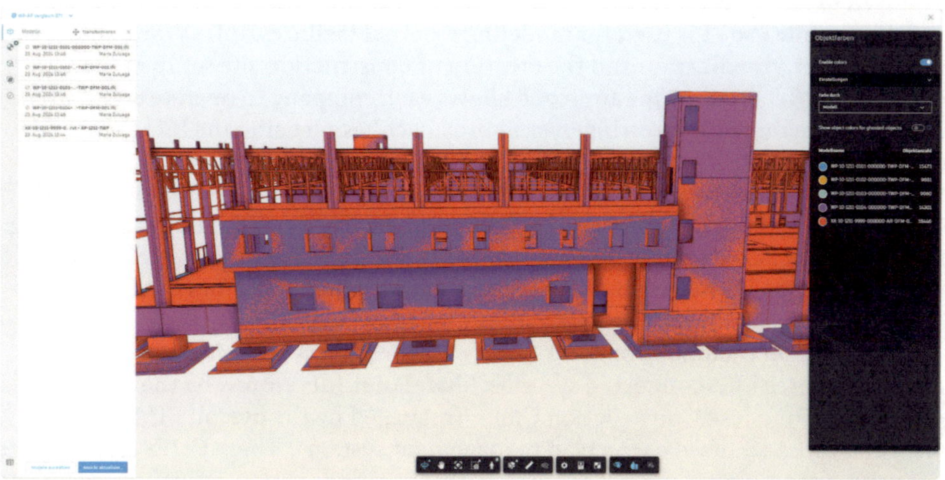

Figure 11.
Visual detection of deviation between models.

approach ensures the effective management of complex workflows, reduces risks, and facilitates collaboration among project participants, eventually contributing to the successful realization of project objectives and establishing a model for future infrastructure projects.

4.3 OpenBIM standards and tools

To ensure continuous collaboration among the various companies involved in the "Neues Werk Cottbus" project, a comprehensive set of guidelines and standards was established from the beginning. This aligns with the DB-BIM Strategy document [7], which emphasizes six key areas for successful BIM implementation: strategy and framework, BIM applications, processes and standards, data and information, IT infrastructure, and communication. The creation of essential documents, such as the BIM Execution Plan (BAP), project information structure, and modeling guidelines, reflects these principles, providing a clear framework for all participants and ensuring alignment toward the successful completion of the project.

The project's complexity is amplified by the involvement of numerous companies, each with its own software systems and standards. The diversity of software tools employed across different firms is extensive; some solutions are shared among multiple companies, leading to overlaps, while others are tailored to specific organizations and their specialized tasks. This diversity requires meticulous planning and coordination to ensure that all participants can collaborate efficiently.

The BIM Execution Plan (BAP) and associated documentation play a crucial role in facilitating this collaboration. However, beyond these documents, a thorough understanding of the entire software ecosystem is equally important. Recognizing this, extensive technical discussions were held to develop a project-specific software structure that selects the most effective tools for ensuring efficient, transparent, and straightforward collaboration among project partners. Each discipline within the project uses software tailored to its specific requirements. For example, Revit is employed for technical building systems, structural engineering, and architectural modeling, while ProVI is used for modeling external facilities. InfraWorks and Desite are utilized for visualization and the creation of construction site setup models, including the 4D model. This approach allows each company to operate efficiently within its domain while ensuring overall project cohesion using the IFC exchange format (**Figure 12**).

To address issues and manage deviations, Autodesk Construction Cloud (ACC) is primarily used, though some planners choose to manage deviations internally using Dalux or BIMcollab Zoom. Scheduling and 4D simulations are facilitated through a combination of MS Project and Desite, providing robust tools for managing the project's timeline and visualizing progress.

The Common Data Environment (CDE) for the project is a critical component, comprising three interconnected systems: SharePoint (developed by the general contractor), Autodesk Construction Cloud (managed by the overall BIM coordinators), and EPLASS (used as the plan management system). These CDEs are integrated

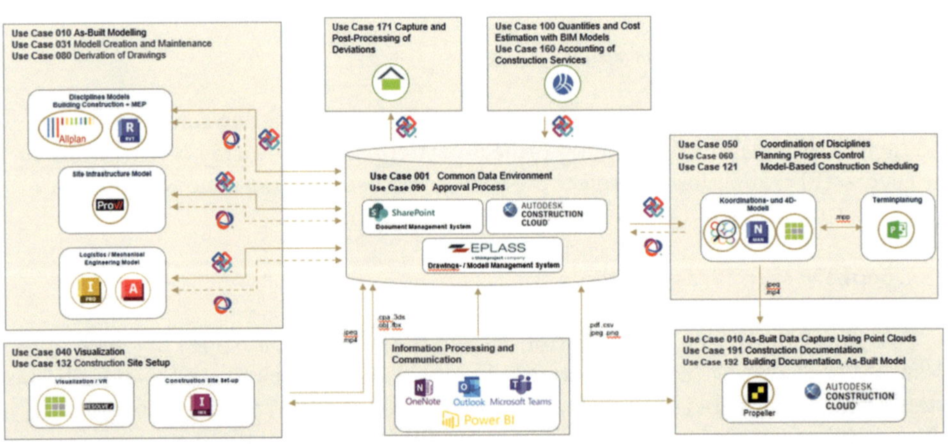

Figure 12.
Project-specific software ecosystem map.

through ACC, ensuring that all data and documents are synchronized and accessible to all relevant parties. This integration is vital for maintaining consistency and transparency across the project.

By establishing these Open BIM standards and carefully selecting the appropriate tools, the project framework supports effective communication and collaboration across all phases of "Neues Werk Cottbus." This structured approach ensures that all participants work together efficiently, contributing to the successful completion of the project.

4.4 Benefits and challenges

The integration of BIM into the "Neues Werk Cottbus" project was a strategic decision aimed at improving collaboration, efficiency, and precision throughout the construction process. While BIM brought substantial advantages, its implementation also presented unique challenges, especially given the scale and complexity of the project. This section outlines both the significant benefits and the challenges encountered during the project's execution (**Figure 13**).

One of the most notable benefits of using BIM in the project was the geometric integration of various discipline-specific models. By aligning all models with the project's central reference point, BIM ensured that every element was accurately positioned, reducing the risk of discrepancies and errors.

BIM also greatly enhanced the understanding of 2D plans by translating them into 3D models. This shift allowed all stakeholders, including those less familiar with technical drawings, to visualize the design in three dimensions and understand them better.

Another significant benefit was the use of 3D models for work preparation. By simulating construction processes within a 3D environment, planners could anticipate challenges and optimize the sequence of construction activities. BIM tools, like Common Data Environment (CDE), were highlighted for their ability to manage

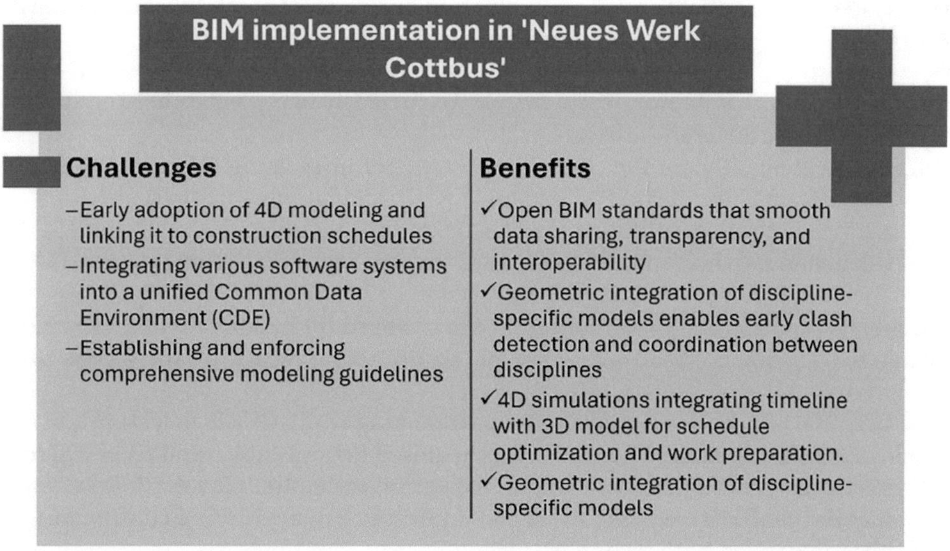

Figure 13.
BIM challenges and benefits overview.

project data efficiently, reducing information loss and facilitating smoother workflow across teams [8]. The use of 4D simulations, which integrated the project timeline with the 3D model, was another critical advantage. These simulations allowed the project team to visualize the construction phases over time, helping to optimize schedules and resolve design problems early.

The use of 4D simulations, which integrated the project timeline with the 3D model, was another critical advantage. These simulations allowed the project team to visualize the construction phases over time, identify potential clashes between different elements or working spaces, and optimize the construction schedule accordingly.

Finally, the client's mandate for openBIM from the beginning played a vital role in ensuring the project's success. By adhering to openBIM standards, the project fostered a collaborative environment where data could be shared smoothly across different software platforms. This approach not only improved transparency and interoperability but also ensured that critical deadlines could be met.

Despite these substantial benefits, the implementation of BIM in the "Neues Werk Cottbus" project also presented several challenges. One of the primary difficulties was the need to establish comprehensive modeling guidelines that could be consistently applied across all disciplines. Ensuring that all project participants adhered to these guidelines was essential for maintaining uniformity in the models, but it required significant effort to coordinate and enforce these standards across different teams and software systems.

Another challenge was the necessity of integrating numerous software systems into a single Common Data Environment (CDE) for managing and sharing information. With multiple companies involved, each using its own preferred software, creating a unified platform was a complex task. Additionally, the use of three different CDEs presented further challenges in organizing information, as it was sometimes unclear where specific data should be stored. This lack of clarity occasionally hindered efficient communication and data management. For future projects, it is crucial to consider the possibility of consolidating the project into a single CDE to streamline the organization and ensure continuous access to information. The integration of these systems is essential to avoid data silos and guarantee that all stakeholders have access to the most up-to-date information, but it demands careful planning and ongoing coordination. The experience highlights how establishing an effective CDE can mitigate issues with data storage and access, particularly in projects where multiple organizations use different software [8].

The project highlighted the critical importance of integrating 4D modeling from the outset. Establishing clear requirements for 4D simulations early in the project was essential to effectively linking the construction schedule with the 3D models. However, achieving this required a strong commitment from all parties to consistently meet these requirements, which proved to be a significant challenge. This difficulty was particularly pronounced due to the need for precise attribution within the models, making accurate data input and coordination crucial for the success of the simulations.

Finally, the project faced challenges in conducting regular clash detection and coordinating the schedule. These activities required not only advanced technical tools but also effective communication among the various disciplines involved. Ensuring that potential conflicts were identified and addressed promptly was a continuous effort that demanded both technological and human coordination.

In conclusion, while the implementation of BIM in the "Neues Werk Cottbus" project presented several challenges, particularly in terms of standardization, platform integration, and early adoption of 4D modeling, the benefits it provided were substantial. The use of BIM not only improved geometric integration and understanding of the design but also enhanced work preparation and coordination. These benefits contributed significantly to the successful execution of the project, demonstrating the value of BIM in managing large-scale, complex construction projects.

5. Added value and synergies of IPD and BIM

The strategic implementation plan for digital design and construction by the German Federal Government defines BIM as follows:

"Building Information Modelling refers to a cooperative working methodology with which the information and data relevant to the life cycle of a building are consistently recorded, managed and exchanged in transparent communication between the parties involved or made available for further processing on the basis of digital models of a building" [9].

An IPA contract, on the other hand, offers the opportunity to rethink traditional processes and overcome the usual interface problems between clients, planners, and construction companies due to the early involvement of all stakeholders in the design and construction process in a joint contract.

Both approaches are primarily concerned with better collaboration in project management based on clearly defined processes. The BIM methodology provides a clean data basis that is available to everyone, with better and easier-to-understand quality based on 3D models. The IPA contract model makes it possible for those people with the necessary knowledge to be involved in the process right from the start of the project.

Specific examples from the Neues Werk Cottbus project will underpin these considerations below.

5.1 Connecting individual experts for new solutions

Within individual companies, BIM experts are often left to their own devices. There are small development teams that only ever look at one part of the entire life cycle of a project, whether from the perspective of the client, a planner, or a contractor. This results in one-sided knowledge that does not do justice to the holistic approach of the BIM methodology.

In the Neues Werk Cottbus project, a joint team consisting of experts from all partners was set up at the beginning: those involved from the client, designers, and construction companies, from shell construction to the technical building equipment to the railway systems, from technology to cost calculation. Each expert contributed their knowledge. This enabled gaps in knowledge and project processes to be bridged together and new solutions to be developed. Above all, this exchange offered the opportunity from the outset to review the solutions developed alone and to overcome silo thinking.

5.2 Minimizing data and information loss

Both approaches, IPA and BIM, are based on a centralized data and communication platform. The documents, models, and data developed are equally available to all parties involved, while the experience gained from the BIM methodology, the standardized processes for storing, and the status of data as defined in ISO 19650 help with joint collaboration.

In contrast to the classic value creation process, where there are always gaps in the transition between preliminary design, detailed design, and work preparation because only some of the data and documents are passed on, all information is retained here and is constantly being further developed and elaborated. This is helped by the fact that all information is shared and available to everyone, but also that everyone involved is involved right from the start. Because even if the data is available transparently (BIM), every new person who joins has to sift through and interpret this data first, asking the same questions again that the others have already clarified at the beginning, which naturally leads to a loss of time. If a large proportion of project members are involved right from the start due to the IPA contract, this loss of time can be avoided (**Figure 14**).

5.3 Simulation of construction processes

The feasibility and cost-effectiveness of a project are significantly determined by effective planning of the construction process. Good planning allows for the selection of cost-effective construction methods, helping to meet deadlines. According to Mi and Li [10], integrating BIM in the preconstruction planning phase enables efficient simulations and optimized resource allocation, potentially reducing project planning time by 20% and material costs by approximately 15%. The 4D planning of the construction process offers the opportunity to simulate construction methods and processes for the various trades directly on the 3D model and thus to see transparently whether all steps fit together. It is precisely in the construction process that the expertise lies with the construction companies. Only through the early involvement of the construction companies by IPD can the 4D planning achieve the quality that brings added value to everyone, because only they have the knowledge of which materials are available, know the prices, the delivery times, and have the experience in the construction processes. Without the 4D model, on the other hand, the experts

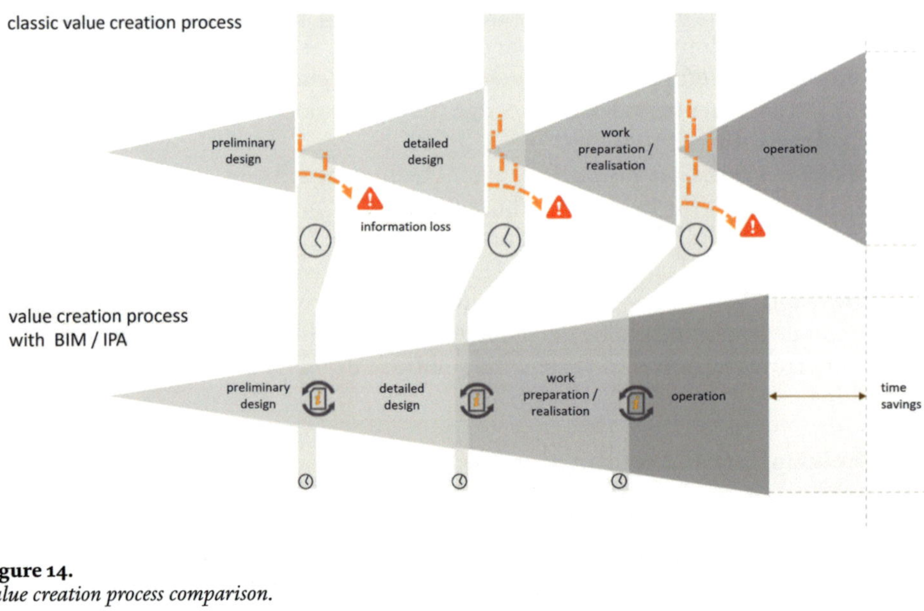

Figure 14.
Value creation process comparison.

involved lack an overview of the spatial relationships that only become recognizable in the model due to a lack of data. Therefore, one of the greatest added values of the combination of BIM and IPD is the optimization of the construction process and construction methods.

5.4 Dealing with new transparency

Mistakes happen in every project and every company; misunderstandings arise, and delays happen. In traditional construction projects, there are many ways in which designers, construction companies, and even clients can hide their own mistakes, using claims and notices of concern or obstruction.

In contrast, the transparency enabled by BIM processes and the close collaboration fostered in IPA projects make any mistakes much more visible. This heightened visibility also brings different working methods and practices into the spotlight, creating a new dynamic that projects must navigate.

Very few people like to admit mistakes, partly because they often fear the unpleasant consequences from previous experiences. To be able to deal with the new transparency, IPA project teams therefore need regular support and coaching to avoid falling into old patterns. A good and appreciative error culture is a decisive factor in the success of an IPA project.

6. Lessons learned and future outlook

The "Neues Werk Cottbus" project offers valuable insights into the evolving application of Building Information Modeling (BIM) and Integrated Project Delivery (IPD), emphasizing the need for continued innovation in interoperability and collaboration. The lessons learned from this project can guide future initiatives toward more efficient and integrated outcomes.

A key takeaway from the project is the importance for early and comprehensive integration of BIM into the project framework. To fully exploit the potential of BIM, it must not remain confined to a specialized group of "BIM experts." Instead, all stakeholders—from project managers to on-site workers—must engage with BIM processes. This will ensure that its benefits are maximized throughout the project lifecycle, contributing to better decision-making, coordination, and overall project success.

However, achieving widespread engagement necessitates significant effort in design and coordination. In large, multi-disciplinary projects, early phases require extensive discussions on the project's information structure, workflows, software selection, and training needs. These first steps are crucial for establishing a shared understanding and aligning all stakeholders with common objectives. To support this process, there is an urgent need for well-defined, standardized guidelines that outline best practices for BIM implementation, the effective use of Common Data Environments (CDEs), and long-term data management for future facility operations.

In this context, the client's role in setting standards is particularly important. Project owners need to establish early guidelines for the types of data that will be used during operations, as well as the CDE platforms and workflows that will be used throughout the project lifecycle. In the future, industry-wide standards will be essential, not only because not all clients have the resources to develop their own standards but also to ensure that designers and contractors can work confidently under consistent principles across different projects and clients.

In this respect, BuildingSMART and other industry associations have a central role to play in promoting cross-industry standards that can be applied to future projects. As a major client, Deutsche Bahn also has an important responsibility in this regard.

Additionally, ongoing updates to project documentation will be crucial. A short manual or guide for all project participants, regularly updated throughout the project lifecycle, can provide clarity on changing processes, standards, and expectations. Such documentation would serve as a reference tool for all those involved, promote coordination between the teams, and avoid misunderstandings.

For future projects, work processes should also be re-evaluated to incorporate BIM methodologies. While the contents of the German HOAI (Honorarordnung für Architekten und Ingenieure) and its outline of various project phases are still relevant, the differing phases in an IPA contract necessitate a careful integration of BIM methodology. Here, it will be important to create new quality gates that ensure the optimum time when the construction partners should be involved and obliged to provide input and when BIM processes are implemented. It must be clear that several iterative steps are sometimes required to achieve optimal results. Defining such an approach at the start of the project helps the teams to overcome unexpected challenges and optimize collaboration throughout the project.

For projects with tight schedules, detailed process simulations will be indispensable. These simulations must engage all stakeholders from the beginning, allowing for a comprehensive understanding of the project's timeline and interdependencies. The early integration of simulation tools helps to anticipate potential delays, refine scheduling, and ensure smoother execution.

Finally, the importance of a single, integrated CDE cannot be overstated. As demonstrated in the "Neues Werk Cottbus" project, the use of multiple CDEs can create confusion and hinder effective data management. Moving forward, the implementation of a unified CDE from the start will be critical for reducing complexity, ensuring that information is easily accessible, and minimizing the risk of data loss or misinterpretation.

In summary, the future application of BIM and IPD methodologies will focus on deeper integration, clearer standards, and more collaborative workflows. Interoperability, supported by industry-standard formats such as IFC and BCF, will remain a top priority to ensure smooth data exchange across all platforms. As the industry continues to evolve, ongoing collaboration with software providers, project teams, and regulatory bodies will be essential to developing new solutions that enhance project outcomes and drive the future of construction.

Author details

Daniel Krause[1*], Claudia Szargan[2] and Maria Zuluaga[1]

1 Wayss & Freytag Ingenieurbau AG, Frankfurt, Germany

2 DB Fahrzeuginstandhaltung GmbH, Taucha, Germany

*Address all correspondence to: daniel.krause@wf-ib.de

References

[1] Was ist IPA—Integrierte Projektabwicklung mit Mehrparteienvertrag? [Internet]. Available from: https://www.dabonline.de/recht/was-ist-ipa-integrierte-projektabwicklung-mehrparteienvertrag/ [Accessed: September 11, 2024]

[2] Rodde N, Boldt A. IPA—Integrierte Projektabwicklung in der Praxis. München: Verlag C.H. Beck; 2024

[3] BLB NRW. Integrierte Projektabwicklung mit Mehrparteienvertrag [Internet]. Available from: https://www.blb.nrw.de/service/service-fuer-auftragnehmer/integrierte-projektabwicklung-mit-mehrparteienvertrag [Accessed: September 11, 2024]

[4] Dabonline. IPA—Integrierte Projektabwicklung: Vorteile, Nachteile, Risiken und Haftung [Internet]. Available from: https://www.dabonline.de/recht/ipa-integrierte-projaktabwicklung-mehrparteienvertrag-vorteile-nachteile-risiken-haftung/ [Accessed: September 11, 2024]

[5] buildingSMART International. Use Case Management (UCM) [Internet]. Available from: https://ucm.buildingsmart.org [Accessed: September 11, 2024]

[6] buildingSMART International. Use Case Documentation V4.1 [Internet]. Available from: https://www.buildingsmart.org/wp-content/uploads/2019/02/bSI-Awards-2019-Use-Case-Documentation-V4-1.xlsx [Accessed: September 5, 2024]

[7] Deutsche Bahn AG. BIM-Strategie der Deutschen Bahn für die Eisenbahninfrastruktur [Internet]. p. 14. Available from: https://www.deutschebahn.com/de/konzern/bahnwelt/bauen_bahn/BIM-6875938 [Accessed: September 5, 2024]

[8] Dalui S, Kumar A, Singh V. BIM and IPD integration: Benefits and challenges in the UK infrastructure sector. Journal of Information Technology in Construction (ITcon). 2021;**26**:930-932. Available from: https://c333711b77c1f4202d7d4ea9201ca0b39407.pdf(semanticscholar.org) [Accessed: October 2, 2024]

[9] Bundesministerium für Digitales und Verkehr. Stufenplan Digitales Bauen [Internet]. Available from: https://www.bmdv.bund.de/SharedDocs/DE/Publikationen/DG/stufenplan-digitales-bauen.pdf?__blob=publicationFile [Accessed: September 11, 2024]

[10] Mi Z, Li J. Maximizing Project Efficiency and Collaboration in Construction Management through Building Information Modeling (BIM). Toronto Metropolitan University; Royal College of art. 2021. Available from: https://(PDF)Maximizing.project.efficiency.and.collaboration.in.construction.management.through.building.information.modeling.(BIM)(researchgate.net) [Accessed: October 2, 2024]

Chapter 4

Examination of BIM and Virtual Reality Technology in the Building Production Process through User Experience

Gülce Bozkurt, Selahattin Ersoy and Enes Yaşa

Abstract

Construction industry is open to innovation and change. The evolution of information technologies led these technologies to be integrated also into building production processes, and it is aimed to carry out the building production processes in a faster, systematic and controlled manner. In this context, Building Information Modeling (BIM) and Virtual Reality (VR) technologies, which are information technologies used in building production processes, are widely encountered. Many conveniences are provided in building production processes in integrated projects carried out with BIM processes and VR tools. The study aims to identify the objectives, strengths and weaknesses of using BIM and VR integration in building production processes and to evaluate them in the context of potential threats and opportunities. A literature review and a company survey based on user experiences have been conducted in this study.

Keywords: building information modeling, virtual reality, building production process, user experience, strengths and weaknesses

1. Introduction

Building production is defined as the activities related to the process of the existence of structures such as buildings, roads, bridges, dams, harbors, etc. to meet the demand or requirement of a customer or a user with certain engineering applications [1]. The building production process, on the other hand, is defined as a system in which the resources brought together for a specific purpose follow a process that includes more than one sub-purpose and action, including the necessary procurement and utilization methods to obtain the planned structure, and ultimately the purpose of creating the structure as a product is realized [2].

The building production process starts with a need or a need of the user that turns into a demand. The aim of the building production process is to create a physical structure. For this purpose, first planning, then design and implementation and finally utilization stages are passed to create the targeted structure. Any problem or obstacle encountered in these stages results in some changes in other stages, which makes the stages interactive and feedback.

Technology has gained a place in every sector due to the superiority and convenience it provides and has adapted to the dynamic and variable structure of the construction sector. In the building production process, many stakeholders from different professions share information, which leads to a large variety and amount of information. To ensure the integrity of this information, the need for information technologies in the construction industry has increased [3].

Today, although there are still those who use traditional methods in the building production process, the use of information technologies is becoming a necessity. Building production is a dynamic and variable formation by nature. With this dynamism in building production processes, information technologies have gained a place in the construction sector with the demands of users and entrepreneurs. Information technologies that are frequently used in the construction industry are Computer-Aided Design (CAD), Project Management and Programming Software, Engineering Analysis Programs, Project Management Software, Building Information Modeling, Virtual Reality Applications, Animations, 3D Product Modeling and Visualization Tools, Geographic Information Systems, Electronic Data Interchange, Mobile Communication Technologies, Document Management, Computer-Aided Business Management, Simulations and Analyses, etc.

Technological developments in building production can be considered more of a necessity than a competitiveness policy. Adopting and utilizing new technologies requires some significant investments. Factors such as need, demand and competition lead to increased investments in information technologies.

Building Information Modeling (BIM), one of the information technologies used in building production processes, is defined as a system that brings together data from multiple disciplines structured to create a digital representation of an asset throughout its life cycle from planning and design to construction and operation. It combines data from different disciplines to digitize it on an open cloud platform for real-time interdisciplinary collaboration. The aim here is to create a 3D model to make better decisions in the building production process and to create sustainable and long-term solutions [4].

Virtual reality (VR), one of the information technologies used in building production processes, expands the boundaries of traditional perception, making expensive and dangerous conditions safe and giving users the freedom to simulate these conditions realistically [5]. In building production processes, these scenarios provide different experiences for customers, work safety simulations or quality control for employees and privileges such as mastering the details for project stakeholders.

Building production processes carried out in integration with BIM processes and virtual reality tools will be more planned, controlled, systematic and economical. Especially in many studies conducted abroad, systems developed with BIM and virtual reality technologies have started to replace traditional design, production and management processes. The solutions produced with this integration open new horizons for the construction industry and are moving toward becoming the new traditional system in traditional building production processes.

The aim of this study is to examine the effects of BIM (Building Information Modeling) and VR (Virtual Reality) integration from information technologies on building production processes with the developing technology in the construction sector. It is aimed to examine the strengths and weaknesses of BIM and VR integration in design, construction and post-construction processes of building production processes and to evaluate potential opportunities and threats. This study has been conducted with the idea that it will be both a contribution to the literature and a guide and informative for offices, company owners or individual employees in the sector who have the idea of digitalization in the sector.

In the first part of the study, BIM and VR technologies were examined by conducting a systematic literature review in the first part as a methodology, and the purposes of the integration of BIM-VR technologies in the design, construction and post-construction stages of the building production processes and their contributions to these stages were accessed. In the second part, case studies in the literature are examined in depth and the purposes of using BIM-VR technologies and the benefits provided by this integration are accessed. The second part is supported by a study conducted with selected companies to support the literature. In the third section, the findings obtained are evaluated and recommendations are made for researchers and the sector in the conclusion section. In this study, which is intended to contribute to the literature and the sector, the company study conducted in the context of transferring the user experiences of the leading companies in the sector is of great importance.

2. Methodology

Within the scope of this study, a systematic literature review on BIM and VR technologies has been conducted, what these technologies are, what they are used for and their usage areas have been examined.

Using VOSViewer software related to the studies on BIM-VR, the connection of keywords in the publications scanned in Scopus in the subject areas of "Engineering," "Environmental Science" and "Arts and Humanities" related to "BIM and VR." is examined. It is observed that the keywords of safety, occupational safety, design evaluation, cultural heritage, engineering education and construction industry in BIM and VR issues are concentrated in the studies for 2021–2022 and appear as new fields of study. It can be said that BIM-VR keywords are related to many keywords such as architecture, engineering, construction, simulation and digital construction and studies are intensively carried out in these areas.

Again in the VOSviewer software, when we look at the countries where the publications scanned in the subject areas of "Engineering," "Environmental Science" and "Arts and Humanities" related to BIM and VR in Scopus are made, it can be said that studies in this field are intensified in countries such as China, the United States and England and as of 2020, Turkey is also involved in the studies on this subject.

Following the literature analysis, the use of BIM and VR integration in building production processes—design, cultural heritage, construction and post-construction phases—was investigated. In the literature study, the existing studies on the subject were examined, and five publications on a case study of BIM-VR integration were examined in detail. As a result of the reviews, the usage purposes, strengths and weaknesses of BIM-VR integration were identified.

As a method that will distinguish this study from other studies, user comments and blogs of companies that offer software related to the integration of virtual reality and augmented reality technologies into the construction sector were browsed. Interviews and interviews with construction companies in user comments and blogs were examined, and their comments on BIM-VR integration were accessed by analyzing the comments. Construction companies using this integration were accessed from the user comments section of the scanned software companies. Case studies, blogs and articles about BIM and VR on the websites of the accessed construction companies were browsed. In light of the findings, the purpose for which construction companies use BIM-VR integration, their strengths and weaknesses were identified.

3. The use of BIM and virtual reality technologies in building production processes

BIM can be briefly defined as a process that aims to collaborate and share information with the project participants from the design phase to the demolition phase of the project, which is aimed to be created with various hardware and software. It is possible to group the usage areas of BIM under four main headings as used in design processes, environmental analysis, building construction processes and building operation processes [6].

VR technology is an environment consisting of computer simulations in which users will receive both physical and psychological feedback in response to these movements by perceiving the movements of the users.

Today, VR technology, which is widely used in many commercial areas ranging from desktop applications to advanced immersive experiences, is offered to the use of building production processes in the construction industry for different disciplines and goals thanks to its simulation and visualization capabilities [7]. The interoperability of BIM and VR can enrich BIM data and strengthen the production, planning and accuracy of the BIM model at various stages of the project life cycle [8]. The integration of BIM and VR technologies has a great potential to contribute to many aspects of building production processes, such as early detection of problems, improvement of design quality and inspection of design components in an immersive virtual environment [9]. Building production processes integrated with BIM processes and VR tools will be more planned, controlled, systematic and cost-efficient [10, 11].

In their publication, Alizadehsalehi et al. [12] summarized the workflow for converting a BIM model into a viewable VR model as shown in **Figure 1**. This workflow starts with the creation of an information-rich 2D/3D-BIM parametric model of the project team including architects and engineers with Revit 2019 software. The BIM 360 cloud server (database) was used to provide an online storage space that all team members can access in real time. To convert the created Revit model into a VR model, the Fuzor plugin in Revit 2019 was used to transfer the model to VR devices such as Oculus Rift, HTC Vive and Samsung HMD. Thus, an immersive BIM/VR adventure was experienced [12].

In summary, as shown in **Figure 2**, the model created by project teams with BIM is exported with a middleware program or plugins offered by technology companies, directed to VR tools and used in building production processes [7]. It is possible to examine the BIM-VR integration used in building production processes in the construction industry, in design, construction and post-construction stages and cultural heritage buildings.

3.1 Using BIM-VR in the design phase

BIM-VR technologies can be used in architectural design stages, working on a single model, visualization, pre-detection of design flaws and conflicts, experiencing the space, making appropriate designs according to emergency plans and many more.

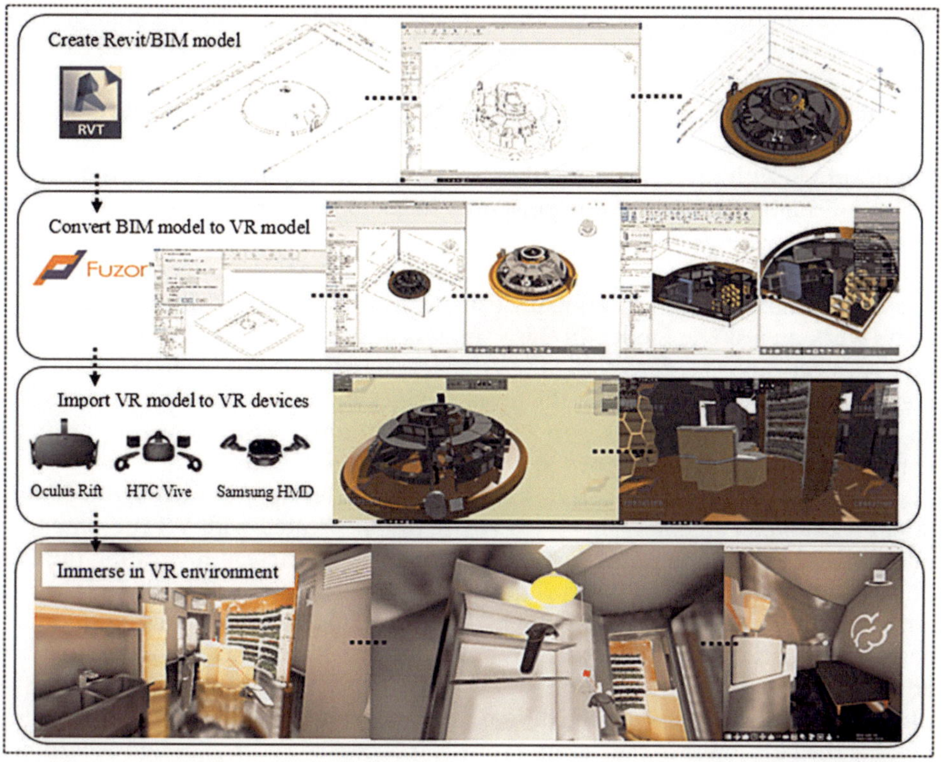

Figure 1.
Transfer of BIM model and VR tools [12].

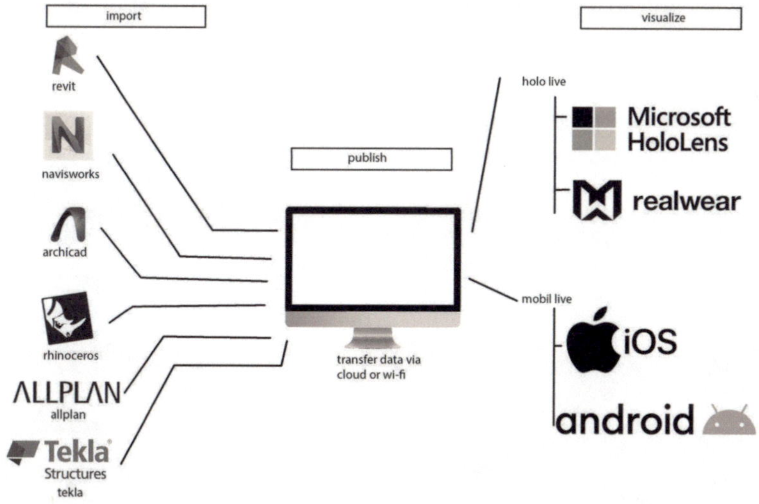

Figure 2.
Transfer of BIM and VR tools [13–16].

The most advantageous part of BIM-VR integration in design processes is that all project stakeholders can work together on the same model. In this way, even if they are in different locations, project stakeholders can carry out the design and drawing of projects together on a single model, and cooperation and communication in the project are at a very high level. Experiencing the project drawn with a VR headset in the background while drawing in a BIM program, as in **Figure 3**, provides the ability to see the details of the project, detect design defects in advance and provide double-sided control [17].

In meetings with project stakeholders, the ability to instantly save the renewed parts and visualize multiple designs instantly helps to save time and increase satisfaction in design meetings [17].

During the architectural design stages, it helps to see MEP projects such as static, electrical, plumbing and ventilation on a single project by having all project stakeholders work on a single model. In this way, it is possible to recognize many errors such as project slippage, detection of overlaps, design flaws, etc., as in traditional systems, with system warnings at the project stage, and to see these details or problems in 3D with VR tools. Thus, cost and time-consuming applications such as returns and reconstruction in building production processes are minimized.

It is possible to use of BIM-VR technologies in a different area during the design phase, for example, an emergency such as fire can be created with VR equipment in a model created in BIM environment. Determining the deficiencies by testing the tools such as smoke detectors and fire sprinkler systems designed during a fire in a virtual environment will help to understand which areas should be paid the most attention in emergency situations on the model. Thus, emergency planning of the building can be made even at the design stage [19].

In the relationship between the design team and the end users/customers, by involving the customers in the design process, it becomes easier to design according to the expectations of the customer in terms of accessibility in the interior and exterior spaces of the building and the functionality of the space. The ability of customers to see what is designed with VR tools in advance reduces feedback, saves time and cost and shortens the initial design phases. It also improves collaboration and

Figure 3.
Design experience with BIM-VR integration during the design phase [18].

communication between the design team and clients [20]. In addition to all these, when the project phase is over, customers can experience the space in 3D and interactively by using instantly accessible visuals, giving a sense of depth and various presentation methods with VR technology. In this way, it will allow sales to be made before the construction site process and will even lead to a noticeable increase in sales.

3.2 Using BIM-VR in building production phase

It is possible to examine the use of BIM and VR technologies on the construction site in three stages: visualization and control stages in the building production process, field organization stage and occupational safety processes.

The biggest advantage of accessing 3D visuals in the use of BIM and VR technologies on the construction site is the ease of understanding the project in real terms. In this way, the building can be easily perceived by the construction site team with a view of the targeted building from different perspectives. Instead of interpreting 2D drawings during the construction process, being able to see the targeted production in 3D with VR tools reduces misunderstandings and helps to strengthen communication. The ease of seeing the production to be made on a 1/1 scale will reduce returns. In addition, the project, which can be viewed on a 1/1 scale, will facilitate the early detection of design flaws and allow MEP and conflict checks as shown in **Figure 4** [21].

In the field organization of BIM-VR technologies, it can be used in the formation of field logistics plans by making and visualizing the layout plan of the construction site. It helps to make all planning such as crane installation plans, storage area of materials and transportation plans of materials within the site and simulate them in 3D. It allows risk plans to be made by performing different scenarios and conflict tests that may occur in the field. In this way, the site organization will be built and planned in 3D against the current situation and scenarios that may occur [23]. The fact that BIM-VR technology offers solutions to support mutual information sharing among project stakeholders enables the uncovering, formalization and integration of information through the mutual sharing of teams such as workers, planners and managers among such project stakeholders [7].

The use of BIM and VR technologies in occupational safety planning involves making occupational safety plans in models created with BIM, which helps to ensure

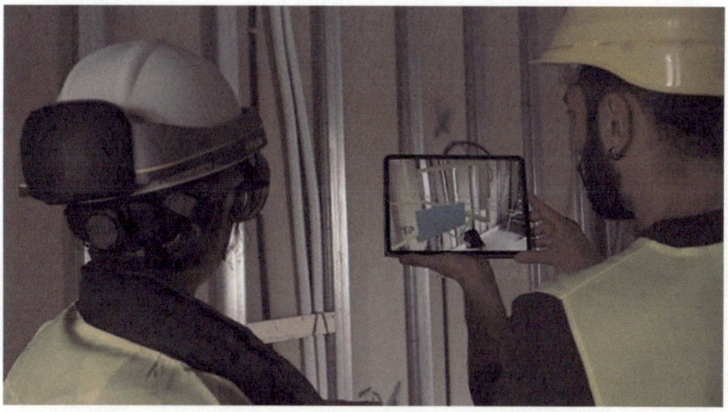

Figure 4.
MEP controls with BIM-VR integration during the building production phase [22].

the correct logistics of the necessary occupational safety equipment for both field personnel and the construction site. With the addition of VR technologies to the BIM system, safe trainings can be provided by using visualization tools in the training of workers and field personnel and field training of trainee students as shown in **Figure 5** [19]. This shows that the combination of BIM and VR technologies will help to reduce worksite safety threats and increase knowledge sharing by identifying safety hazards in advance and training field workers. As a result, language barriers can be removed in a multilingual environment, helping to improve the overall health and safety environment in the field [24].

3.3 Use of BIM-VR in the post-building phase

The most long-term process from the existence of a building to its destruction is the facility management process of that building. Facility management and operations, which is the 7th dimension of BIM's parameters, deals with the operation and maintenance of the building throughout its life cycle. In addition, thanks to the integration of BIM with simulation models and VR equipment, instant access to data can be provided and decisions to be taken about the facility can be helped. While in the facility, with the help of a VR tool, you can see the warnings of the team that previously accessed the model or send a warning message to the team in that area [26]. Renovations can be carried out with MEP controls without problems in accessing main lines and without material losses, and demolition projects can be planned and supported by simulations as seen in **Figure 6**.

Figure 5.
Field trainings with BIM-VR integration during the building production phase [25].

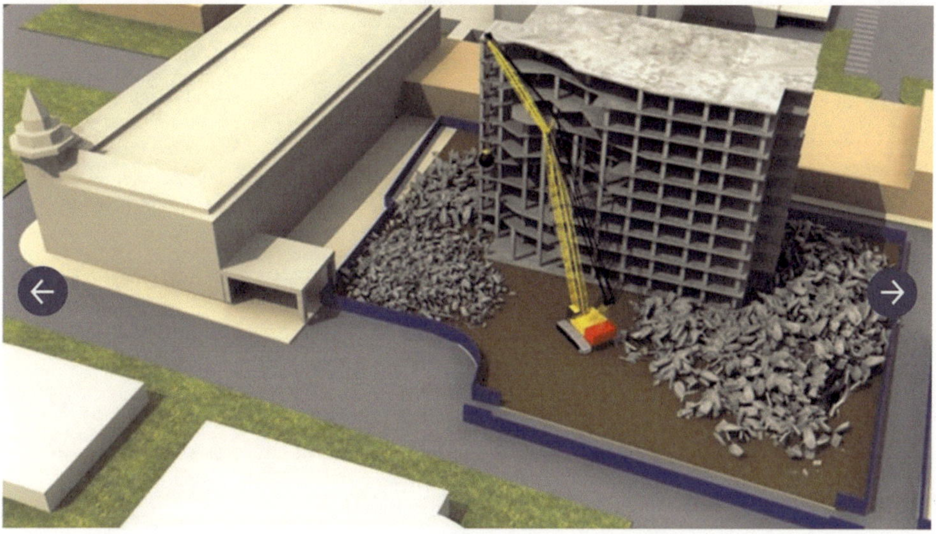

Figure 6.
Building demolition planning with BIM-VR integration in the post-construction phase [27].

4. Case studies and company studies examining BIM and virtual reality technology in the building production process

Two methods of analysis were followed: case studies of existing academic studies on the use of BIM and VR technologies in building production processes and research of software construction companies. Five case studies, where BIM and VR Technology were used, were analyzed (**Table 1**).

4.1 BIM-VR review on construction companies

The method of analysis that can distinguish this study from other studies is the study in which sample software companies and construction companies involved in building production processes with BIM-VR integration are investigated.

In the study, five leading software companies in the sector were scanned, the first of which is Unity, a monopolized company in the gaming industry. In addition to the gaming and film sectors, Unity is also involved in the automotive, energy, aviation, architecture and engineering sectors, and was chosen because it is thought to contribute to the study as a globally recognized company with its usage areas such as digital twins, education and guidance, design, sales and marketing, simulation and human-machine interface. The second software company analyzed is IrısVr/ Prospect, which has made a brand as the number one VR platform in the construction industry. IrisVr software company was preferred because it is an award-winning VR company with more than 60,000 construction companies, making design and model reviews fast, accurate and collaborative. VrCollab was chosen for its global contribution to the study as it has centers in China, India, South Korea, Japan, Hong Kong, the United Arab Emirates, Saudi Arabia and Brazil. The fourth company examined was The Wild. The reasons for analyzing this company can be listed as the ease of integrating files to facilitate AEC flows, offering remote collaboration and accessing metadata in BIM files. But the most important reason is Autodesk's acquisition of The Wild, an

Publication	Purpose	Strengths of Using BIM-VR	Weaknesses of Using BIM-VR
[28]	1. Risk Planning 2. Field Planning 3. Occupational Safety Planning 4. Emergency Planning 5. Prevention of Work Accidents 6. Identification of Potential Hazards	1. Strengthening Communication with Employees 2. Seeing Occupational Safety Equipment in 3D 3. Simulations Simplify Setup 4. Prevention of Work Accidents 5. Occupational Safety Planning and Training 6. Field Planning 7. Effective and Interactive Field Management 8. Foreseeing Potential Threats	1. Calibration Problems 2. Start-up Costs 3. Detailed Preliminary Study 4. Lack of Motivation
[29]	1. Creating Project and Model for Restoration Project 2. Single Project Stakeholders Combining in a Model	1. Ensuring the Continuity of Cultural Tourism 2. Documenting Historical Studies 3. Online Cultural Heritage Visits	1. Detailed Preliminary Study
[30]	1. Ensuring MEP Controls 2. Review of Architectural Design	1. 360-degree angles make it easy to see MEP connections that are not visible in the field 2. Practicality and Speed of VR Use 3. Giving the Feeling of Being Really Inside the Building 4. Strengthening Communication between Teams 5. Strengthening Presentation to the Customer	1. Length of the Acclimatization Process 2. Physiological Challenges 3. Software and Hardware Costs 4. Adaptation Process with Technology 5. Firms' Resistance to Change
[31]	1. Strengthening Communication 2. Reducing Costs 3. Minimizing Rebuilding and Returns in the Project 4. Capturing the Details	1. Team Management. 2. Cost Savings. 3. Recognizing Design Defects in Advance 4. Ease of Modifications in Facility Management	1. Scarcity of Expert Human Resources 2. Multi-Participant System Prolongs the Design Process 3. Initial Cost
[32]	1. Uniting Project Stakeholders in a Single Model 2. Getting Feedback from Users 3. Monitoring the Production Process in Advance 4. Project Coordination 5. Minimizing Changes 6. Site Management and Occupational Safety Planning	1. Cost and Time Savings 2. Creating Workflow 3. Easy Quality Control 4. Easy Access to a Single Project for Project Stakeholders 5. Shortening the Design Process 6. Increased Coordination and Cooperation among Stakeholders 7. Customer Satisfaction through Participatory Design	1. Detailed Preliminary Study 2. Lack of Calibration between Stakeholders 3. Initial Costs 4. Rendering Problems 5. Obligation to Create a Project Finished in All Details

Table 1.
Classification of case studies.

immersive technologies software company. The last VR company, Resolvebim, was chosen because it helps to maintain high standards of safety, quality and sustainability in construction and works with globally recognized construction companies. Resolvebim has been developing itself not only in VR but also in AR and has developed software that can be used in XR devices. Case studies were accessed by scanning the blogs on the website, considering that it would contribute to the study with its features such as converting speech to text, enabling easy and multi-user collaboration, annotating by measuring and drawing, having problem tracking integrations and being able to be followed with the desktop assistant application.

The customers, user comments and blogs of the Websites of the five software companies were scanned, their evaluations of the VR technology they use were systematically grouped, their comments and interviews about BIM-VR integration were compiled and the purpose for which they use BIM-VR integration and their usage experiences were accessed. **Table 2** shows the opinions of construction companies on the purposes of using BIM-VR integration.

In the software company research, 16 blog posts, interviews and case studies from Unity were analyzed. In the software company research, a total of 40 studies, 7 of which are comments and 33 of which are blog posts, interviews and case studies, were analyzed from IrısVR company. The website of VRCollab was scanned and two case studies that will contribute to the study were accessed. The Wild company Website was scanned, and nine case studies and interviews were accessed. The Resolvebim Website was searched and six case studies and interviews that will contribute to the study were accessed.

In total, 73 user experience outputs were analyzed. The expression and equivalents of the opinions evaluated as the strengths of BIM-VR integration by construction companies are given in **Table 3**.

According to this, the most mentioned strength of BIM-VR integration of construction companies is customer satisfaction, which was mentioned in 45% of the answers. This is followed by time savings in 43% of the statements, being able to see details with 3D models in 41% of the statements and cost savings in 40%. The least mentioned strength is the advantages in site planning and management and the benefit in providing overlap-MEP controls, which accounted for 9% of the statements.

Accordingly, the most cited strength of BIM-VR integration by construction firms is certainty, savings and efficiency in cost and project time, which was cited in 15.92% of the responses. This is followed by access to a more detailed model with 14.15% of the statements and sharing a common language with 12.38%. The least mentioned strength was the ease of safety and risk planning, which was mentioned in 2.65% of the statements.

The statements and responses of the opinions that are considered weaknesses of BIM-VR integration by construction companies are given in **Table 4**.

Accordingly, the most frequently mentioned weakness of construction companies in BIM-VR integration is the initial costs, which constitute 40% of the answers. This was followed by the cost and duration of the training to be given to the employees and the length of the familiarization process with these technologies, which accounted for 12.5% of the responses. The least mentioned weakness was the long duration of work, which accounted for 7.5% of the responses.

In the continuation of the study, 10 construction companies with global operations were selected from the construction companies reached by scanning the customers, user comments and blogs of the five software companies that were examined. The building types, countries and BIM-VR tools used by the selected construction

Statements in product reviews, blog posts, interviews and case studies	**Correspond to**
1. "To streamline processes and improve overall collaboration" YVR Airport-Director of Innovation and IT 2. "To increase cooperation" Sitowise-Expert 3. "In order to involve subcontractors and design teams" SPACE-Landscape Designer 4. "Foster better communication and understanding in coordination meetings" Gilbane-VDC Director 5. "To ensure optimization of communication and cooperation" Barton Malow-Team Members 6. "… became apparent to me that this was a tremendous opportunity to get the team together and collaborate, share ideas … " Leo A Daly-Director of Design 7. "bring multiple people into one space, enable them to hold collaborative design reviews, and offer a good desktop option" Dillon Consulting-Planning and Design Group	Increased Communication, Cooperation and Coordination
1. "Bypassing the traditional model of spatial analysis" Losci-Ceo/Founder 2. "Developing design-level insights" Zutari-Interactive Visualization Expert 3. With the aim of catching design flaws early" SHOP-AR/VR R&D Leader 4. "Our goal was to better understand the scale relationships between each structure" StudioMB-Lead Product Designer 5. "To support traditional design with drawings and 3D models, to involve designers in the creative process" SHOP Architects-Team Members 6. "… to flesh out ideas and the initial concept stages and to immerse people in the spaces themselves." Contagious-Director of Design 7. "… to build a persistently evolving virtual space that could replace their usual 2D collaboration … " Adidas-Decision Scientist 8. "… to future-proof their services and design smarter." BSA LifeStructures -Architectural Designer	To be able to improve the design, to master the details
1. "Reduce costs" Zutari-Interactive Visualization Expert 2. "To ensure efficient and cost-effective workflows" One Alliance-Team Member 3. "Their goal was to determine if adidas could reshape its ways of working into something new, while realizing a net cost savings, rather than a significant investment and reorganization." Adidas-Decision Scientist	Reducing Costs or Saving Money
1. "Reduce time" Zutari-Interactive Visualization Expert 2. "To solve problems quickly and shorten the construction time" SHOP-AR/VR R&D Leader 3. "We needed a faster way to make design changes" BRC-Creative Director	Saving Time
1. "To address facility challenges such as maintenance, sustainability and security" YVR Airport-Director of Innovation and Informatics	Making Maintenance and Facility Management Systematic

Statements in product reviews, blog posts, interviews and case studies	Correspond to
1. "To customize housing according to individual needs" House by Urban 2. "To improve customer experience" Schnabel Engineering-Partner 3. "We use it to explore material and design options with customers" StudioMB-Team Member 4. "Improving the early stages of visualization" Anglian Water-Team Members 5. "We wanted to have the client see what were dreaming about and thinking about." "... showing them a "close to reality" version of the design concept." LEO A DALY-Team member 6. "We share it with clients as well, allowing them to see the designs and feel how it looks rather than just see it two-dimensionally on the screen." Contagious-Director of Design	Customer Satisfaction and Effective Presentation to the Customer, Visualization
1. "Aim to reduce changes in the field" Anglian Water-Team Members 2. "To reduce site changes" One Alliance-Team Member 3. "Improving workflows" Friis &Moltke-Team Members 4. "The goal of these sessions was to deliver an as-built 3D model that can be used by field teams to plan maintenance before accessing equipment in the field." Nodum-CTIO	Ease in Site Planning and Management
1. "Reducing accidents" "Increasing workers' awareness of safety factors on the construction site" Skanska-Project Leader/Production Manager/Technology Director 2. "With the goal of improving security" Sitowise-Expert 3. "Reducing risks with our goals" Schnabel Engineering-Company Partner 4. "Identifying security threats in a controlled environment" Barton Malow-Project/VDC Engineer 5. "Improving workplace safety is our primary goal" Balfour Construction-Technology Center of Excellence Leader	Occupational Safety and Risk Planning

Table 2.
Construction companies' intentions to use BIM-VR integration.

companies were analyzed. In addition, the blogs of the selected 10 construction companies about BIM and VR technology on their Websites, case studies and evaluations about the use of these technologies were examined, and as a result, the construction companies' purposes of using these technologies and their experiences about these technologies were accessed (**Table 5**).

Table 6 summarizes the data obtained from the projects in which BIM-VR Technology is used when the determinations and blog posts on the case studies on the websites of software and construction companies are scanned.

5. Results and discussion

As a result of the case studies of the existing studies in the literature and the data extracted from the opinions of construction companies on the use of BIM-VR

Testimonials in product reviews, blog posts, interviews and case studies	Correspond to
1. "It's great to see the team collaborating in real time" Mortenson-Emerging Technology Developer and Project Solutions Technical Specialist 2. "A rich material for interaction with stakeholders" Zutari-Interactive Visualization Expert 3. "Enables client and developer teams to collaborate on real-time 3D design reviews" Mortenson-Technology Developers 4. "Ensures good communication with key stakeholders" SHOP-AR/VR R&D Leader 5. "Solves designers' coordination problem" Gilbane-VDC Director 6. "Reduction of decision-making time, from 15 days to 15 minutes" Gilbane-VDC Director 7. "Better collaboration, faster decisions, fewer mistakes" Gilbane-VDC Director 8. "Just being able to collaborate in were seeing in a 1:1 scale helped us to relocate a few items. This made the whole section more constructible in general." A.T. Chadwick-BIM Detailer 9. "Stakeholders are more engaged in the design approval process" "and at the end of the day we are all able to come to a consensus on how to move forward if we have any design issues that holding the team & the project back" Boustead Project-Digital Delivery Manager & Specialist 10. "We use it resolve issues during coordination meetings." CCDC-Head of Corporate VDC/BIM 11. "Compared to just seeing a big red X over a page, with VR you can hear why we got rid of a slide by leaving a voice note." Leo A Daly-Team member 12. "allowed us to utilize a collaborative platform that immersed key stakeholders in an environment with more complete information to work from, as well as a chance to work together to make changes in the same virtual space." Adidas-Decision Scientist 13. *"When you get the hang of it, then I see it as a faster workflow than the traditional way of coordination." Fortis-BIM Manager*	Increased Communication, Cooperation and Coordination
1. "Saves time and money as well as design functionality and builds stakeholders' trust before the construction phase" Losci-Ceo/Founder 2. "Capturing the hallway column issue in VR saved more than $500,000" "We were able to save an estimated $45,000 in the physical model" Mortenson-Emerging Technology Developer and Project Solutions Technical Specialist 3. "Good forecasts have increased confidence in resource planning" YVR Airport-Director of Innovation and Informatics 4. "We were able to stay within budget and schedule without the expense and lead time associated with physical models" Mortenson-Technology Developers 5. "Less field errors and big savings through more accurate documentation and construction management" SHOP-AR/VR R&D Leader 6. "Productivity gains of up to 15–20%, in some cases can be higher" Sitowise-Expert 7. "It took 18 months to build all the floors, traditionally it would have taken 2–3 years" Kane-Digital Construction President 8. "We saved money" Barton Malow-Project/VDC Engineer 9. "€23,000 reconstruction cost savings achieved" Anglian Water-Team Members	Cost and Project Time Certainty, Savings and Efficiency

Testimonials in product reviews, blog posts, interviews and case studies	Correspond to
10. "Saves time and costs as there is no need to create a model" Friis & Moltke-Team Members 11. " ... that we can get that sooner allows us to be more effective at documenting it, continuing with the design, and moving forward." Leo A Daly-Team member 12. "gets us closer to that data faster and in a more risk averse way." ZoomCare-Director of User Experience 13. "Over a six-month period it was identified that conservatively, on a single project, The Wild could help reduce 150 hours of RFI time, equating to $9521.25 in cost savings." BSA LifeStructures-Architectural Designer 14. "It's time-saving ... " RaumZeit 15. "Subcontractors and design consultants have flocked to the VR review sessions because it saves them time in completing their design reviews and sign-offs" "Resolve ultimately minimized construction delays, change orders, and rework, thereby resulting in significant cost savings for the project" Fortis-Senior MEP Project Manager 16. "The VR reviews surfaced over 600 comments from the project team and about 10% of those comments were high impact issues like the ones described above. These items quickly add up to become substantial budget overruns." Mortenson-Integrated Construction Manager 17. "Interxion saved approximately $216,000 USD in just a few months" Digital Realty Schweiz-Director Operations 18. "Resolve gives our ops teams an amazing gain in efficiency that helps us deliver better data centers. When we review 3D models on a screen share we are wasting almost 50 man hours during a 1 hour meeting- and we are all competing for what to focus on. And with 2D drawing reviews I can generate maybe 2 comments in an hour. With Resolve I provide 30–40 comments in 45 minutes." Critical Facility Manager	
1. "It allows us not to iterate on design changes, but to identify and address gaps in the design" Mortenson-Emerging Technology Developer and Project Solutions Technical Specialist 2. "Enabled medical stakeholders to pre-experience and adapt new working environments at scale" Mortenson-Technology Developers 3. "Explore alternative designs" Mortenson-Project Solutions Technologist 4. "Spatial relationships, space volume and basic design features, scale could be discussed" Mortenson-Project Solutions Technologist 5. "Design review became meaningful, we were able to conduct ergonomics and usability tests" SHOP Architects-Team Members 6. "To be able to perceive the design as in real life with sun settings tools" BB + C Arch-Team Member 7. " ... during our design meeting where the designers are engaged to use VR to get their understanding and approval." Boustead Project-Digital Delivery Manager & Specialist 8. " ... one of the really nice features is the back-end code and web-based interactions so I can go into a physical space and I can interact with digital experiences." ZoomCare-Director of User Experience	Improving the Design with Details

Testimonials in product reviews, blog posts, interviews and case studies	Correspond to
9. "It cracks open the design process" **Dillon Consulting-Planning and Design Group** 10. "I cannot imagine producing the shop drawings without reviewing the design through VR." *Fortis-BIM Manager*	
1. "Having everyone in VR at the same time accelerates decision-making" Mortenson-Emerging Technology Developer and Project Solutions Technical Specialist 2. "a virtual model that multiple stakeholders can access simultaneously" Zutari-Interactive Visualization Expert 3. "Being able to instantly export the model from Revit and Rhino to the cloud and live synchronization with multiple remote users. Very impressive" Retima-Architect 4. "Enables client and developer teams to collaborate on real-time 3D design reviews" Mortenson-Technology Developers 5. "Access for all who serve in a unified vision" Vectuel-CEO/ Innovation Director 6. "Ease of remote meetings due to Covid" Gilbane-VDC Director 7. "Successful in remote working and provides the feeling of being in the same room" KPF-VR Expert 8. "It's one thing to just send files to someone to review, but it's an entirely different thing to actually be in the project and point out things that the team can immediately address." A.T. Chadwick-Director of Construction Technology 9. "online meetings are going to change the way we work" . Boustead Project-Digital Delivery Manager & Specialist 10. "There's an ability in The Wild to capture a photograph of the changes, and we can send that out electronically to all team members and incorporate it into the Revit model right away." BSA-Chief Design Officer 11. "As the platform has been updated, we have folded in those new pieces. For example, LIDAR scanning—somebody with an iPhone can capture an object and put it in an environment without having to hire a super expensive consultant and fly them all over." Leo A Daly-Director of Design 12. "we are able to still get multiple people in one space virtually, which is game changing." Contagious-Director of Design 13. "and allows people who do not necessarily have the software or technical skills, but who know what they'd like, to be part of the conversation" Dillon Consulting-Planning and Design Group 14. "Despite being a fully federated hyperscale data center model, the Navisworks file ran flawlessly on the Quest 2 without everyone needing for a gaming PC or large GPU" Mortenson-Integrated Construction Manager	Sharing a Common Language
1. "Removing uncertainty about how spaces will look and feel" Mortenson-Emerging Technology Developer and Project Solutions Technical Specialist 2. "Buyers can interact with the 3D model and perceive space and objects" "Realistic details can be added" House by Urban 3. "To experience the model in more detail" Schnabel Engineering-Company Partner 4. "Can visualize the output" Schnabel Engineering-Company Partner 5. "Ability to get inside designs with a single click" Payette-3D Visualization Manager	See More Detailed Models and Details in 3D

Testimonials in product reviews, blog posts, interviews and case studies	Correspond to
6. "We were able to observe the cramped and narrow spaces in the model" Gilbane-VDC Director 7. "Camera sites could be identified for security" Barton Malow-Project/VDC Engineer 8. "As designers made changes to the 3D model, they could quickly recreate the virtual reality experience with one click" Miele-Sales Manager 9. "It helped us design better by engaging our sense of scale and proportion" Deforest-Manager 10. "When you view the Project in VR, the valves that you thought were accessible are no longer accessible as you once thought. There really is no comparison to viewing the project in 1:1 scale with your stakeholders." A.T. Chadwick-Director of Construction Technology 11. "... even without VR requirement in tenders, we still use it in our daily process. It really helps give us a better understanding of the building." CCDC-Head of Corporate VDC/BIM 12. "... feels better than the alternatives, which stay in a 2d environment." Leo A Daly-Director of Design 13. "I look forward to being in that space because I know when I'm in that space, I'm at my creatively richest moment." ZoomCare-Director of User Experience 14. "... prototyping and design review tools, everyone could test their experiences before they were physically built and uncover potential flaws that would have otherwise been overlooked had they not explored the space in VR." Adidas-Decision Scientist 15. "It's much easier to engage with operators in VR versus on a screen share. They understand the model faster during model deliveries, and we think that the same will happen during maintenance planning" Nodum-CTIO 16. "We did not have to try to understand a plan; we had the feeling that we were already experiencing the finished result" "... participants could move freely, turning in any direction, stepping in closer, or taking a few steps back, just like in a real room" "It did not take long for the first errors to be discovered" Digital Realty Schweiz-Director Operations	
1. "Allowed evaluation of every component of the assembly" Mortenson-Emerging Technology Developer and Project Solutions Technical Specialist 2. "Digitally catching errors before the job site is a big advantage" Stuffolk-Construction Solutions Director 3. "Coordinates MEP projects" Mortenson-Team Members 4. "Hard-to-detect conflicts are immediately recognized when they are right in front of you" Perkins+Wills-Team Member 5. "We use to pull out specific measurements -especially height – and the size of piping. The ability to inspect the data in those elements is huge for us." A.T. Chadwick-Director of Construction Technology 6. "One technician spotted a light fixture that blocked access to electrical feed boxes. If left unnoticed, the team would've received this feedback during a site walk resulting in a $26,500 change order to rip things out." Mortenson-Integrated Construction Manager 7. "... It did not take long for the first errors to be discovered" Digital Realty Schweiz-Director Operations	Ensuring Conflict and MEP Controls

Testimonials in product reviews, blog posts, interviews and case studies	Correspond to
1. "Helps monitor construction and provide planning guidance on when and where components should be installed" Zutari-Interactive Visualization Expert 2. "The most productive and efficient workflows have been achieved." Mortenson-Technology Developers 3. "Less field errors and big savings through more accurate documentation and construction management" SHOP-AR/VR R&D Leader 4. "Being able to simulate construction is the main benefit" Vectuel-CEO/Innovation Director 5. "The ability of construction to offer VR models" Gilbane-VDC Director 6. "Reduces the impact of site constraints" Gilbane-VDC Director 7. "An effective way to move construction sites into the digital world" Gilbane-VDC Director 8. "It enabled me to communicate with the field teams" Barton Malow-Team Members 9. "Ease in identifying and solving problems that arise on the construction site" "Helps to understand the viewpoints of machine operators" One Alliance-Team Member 10. "the owner of the facility has extended the life of their 3D models beyond design and construction and increased their return on investment in those digital assets. Most importantly, however, the owner's investment in the 3D model will continue to compound with quicker maintenance and smoother operations." Nodum-CTIO 11. "The use of Resolve allows teams to confidently assess the site in VR for issues before it gets built and increases engagement with our existing 3D models, improving the quality of the real-world product" Engineering Implementation Manager- @ one Alliance	Ease in Site Planning and Management
1. "We use VR solutions for safety training in offices" Zutari-Interactive Visualization Expert 2. "Increased employee awareness of occupational hazards" "Fewer accidents by creating safer workplaces" "Provides the opportunity to eliminate risks" "Creates awareness that traditional classroom and online training can never do" Skanska-Project Leader/Production Manager/Technology Director 3. "Helping to identify potential hazards and perform VR safety checks" Barton Malow-Team Members	Ease of Occupational Safety and Risk Planning
1. "Updated visualizations with data collected from the field with drones can be sent to customers every week" Zutari-Interactive Visualization Expert 2. "Provides instant customer feedback" Mortenson-Technology Developers 3. "Sales was the highlight of their presentation" Mortenson-VR Developer 4. "One of the most important parts is the user experience as a whole" House by Urban 5. "The virtual tour allows us to quickly adapt to clients' needs and demands in design and even test these changes" Sound Hannam-CEO 6. "We were able to satisfy our customers very early in the project" Kane-Digital Construction President 7. "Clients felt engaged and knowledgeable about the design progress" Reed-Team Members	Increased Sales and Customer Satisfaction

Testimonials in product reviews, blog posts, interviews and case studies	Correspond to
8. one of the most important intangible benefits of using VR collaboration is the peace of mind stakeholders gain through experiencing projects at human scale. BSA LifeStructures-Architectural Designer 9. "immersed in a space with all these images and videos, I then open up my Library and pull out their 3D models—it's the best sales process ever." RaumZeit	
1. "Enhanced customer experience through a faster and smoother 3D program" Marxent-Kurucu 2. "It allows project stakeholders and clients to visualize, review and optimize designs as interactively as they wish" Zutari-Interactive Visualization Expert 3. "Helps clients truly visualize, experience and validate new design projects" Mortenson-Technology Developers 4. "Sales was the highlight of their presentation" Mortenson-VR Developer 5. "The advantage of customers understanding design" Reed Hilderbrand-Senior Member 6. "Our primary visualization tool" Thornton T.-Senior Project Engineer 7. "Provided a strong visual representation" "Users had a more emotional and exciting experience" SPACE-Landscape Designer 8. "It helped us to show the client the color choice and finish pattern" StudioMB-Chief Product Designer 9. "*We use it a lot to showcase our designs to clients.*" CCDC-Head of Corporate VDC/BIM 10. "The team can put clients in the environments so that they can respond to it." BSA-Chief Design Officer "... allows us to immerse users and clients within the project environment so that they can understand what we are talking about." BSA-Architectural Designer 11. "We can get decisions more quickly by immersing the client in this environment." Leo A Daly-Team member 12. "For customers, a virtual tour of the data center is a simple and impressive experience" Digital Realty Schweiz-Director Operations	Effective Presentation and Visualization to the Customer

Table 3.
Statements that construction companies evaluate as strengths of BIM-VR integration and their responses.

integration in building production processes, it is possible to explain the usage objectives of BIM-VR technologies in building production processes as follows:

- The main purpose of BIM-VR integration is to share a common language and strengthen communication with stakeholders.

- In architectural design stages, it is possible to say that it is used to improve the design, to master the details and to create models in restoration projects.

- In the construction process, BIM-VR technologies are used to make project follow-up easier, to achieve speed and quality in production, to reduce costs, to reduce returns and changes, to ensure MEP coordination and to benefit from the convenience it provides in occupational safety, risk and site planning.

Testimonials in product reviews, blog posts, interviews and case studies	Correspond to
1. "Technology investment made" Retima-Architect 2. "Special VR room set up on hotel grounds for field team" Kane-Digital Construction President 3. "We built a technological infrastructure" Stuffolk-Construction Solutions Director 4. "We had to switch from a temporary VR room to a permanent VR room" SPACE-Landscape Designer 5. "VR team formed" "Software installed" Gilbane-VDC Director 6. "We made a technology investment" Gilbane-VDC Director 7. "We initially bought an installation" StudioMB-Chief Product Designer 8. "We bought the installation" Barton Malow-Project/VDC Engineer 9. "We started with the installation of the VR office, we strengthened the technical infrastructure, we put big screens whenever we have the opportunity" SHOP Architects-Team Members 10. "VR lab installed" "We created a room for team members and customers" Deforest-Manager 11. "We tried to somehow influence and convince big investors to invest in the project" Agora Architecture-Chief Architect and Team Members 12. "We had to set aside €5000 for a 2–3 week trial" One Alliance-Team Member 13. "When using VR, the first mistake you find on a Project will easily pay for the entire year's worth of hardware and software." A.T. Chadwick-Director of Construction Technology 14. "You can go set up your board room and have everybody show up in ... " Leo A Daly-Director of Design 15. "A small cost for a big benefit" ZoomCare-Director of User Experience 16. "We're trying to make ourselves more lean, and sometimes being more lean means you have to invest upfront for headsets and software." BSA LifeStructures-Architectural Designer	Start-up Costs
1. "Bypassing the traditional model of spatial analysis" Losci-Ceo/Founder 2. "We noticed a reluctance to use the tool when others were watching and made room changes" StudioMB-Chief Product Designer 3. "Additional commission charged to customers for VR use" SHOP Architects-Team Members 4. "We realized that Revit was not enough for exports" Agora Architecture-Chief Architect and Team Members 5. "Investments have led to additional commissions from clients" Reed-Team Members	Adaptation to Technology
1. "We had to transfer projects" SPACE-Landscape Designer 2. "You have to make the most of the time spent preparing your model, the time spent in VR and the additional benefits" "Do not be distracted by incomplete elements" Perkins+Wills-Team Member 3. "You need to spend time modeling" Reed-Team Members	Long-Term Work
1. "We are looking for solutions to help us visualize all the data, for now we are using the data shell" Schnabel Engineering-Partner 2. "We explained VR software to clients" SPACE-Landscape Designer	Length of the Acclimatization Period

Testimonials in product reviews, blog posts, interviews and case studies	Correspond to
3. "We are trying to set up the new VR office" "Customers think VR is nauseating, so they are hesitant to try it" Gilbane-VDC Director 4. "We constantly checked whether employees could use technology effectively" Reed-Team Members 5. "trying to plug in 3D files from a different piece of software and retrofit" Contagious-Director of Design	
1. "Project team strengthened" Marxent-Founder 2. "We created a Virtual Insight team" Mortenson-Technology Developers 3. "A Technical R&D group was created to develop the new technology" Schnabel Engineering-Company Partner 4. "Information Workstations created" Zebradog-User Experience Lead	Ensuring Human Resources Adapted to Technology
1. "Necessary in-house trainings were provided" Mortenson-VR Developer 2. "We received 1–3 hour trainings" SPACE-Landscape Designer 3. "The whole team was introduced to the new system" Vice President of Treehouse-Creative+Design 4. "We had employees trained" "Our in-company trainings continued as long as there were newcomers" Reed-Team Members	Cost and Duration of Trainings to be Provided to Employees
1. "Most people are reluctant to start VR because they do not see the benefit of an incomplete model" Perkins+Wills-Team Member 2. "We thought that there would be no effect other than the first customer" Agora Architecture-Chief Architect and Team Members 3. "had a tough start with staff members being skeptical about the value it would bring" Fortis-BIM Manager	Reluctance to Technology

Table 4.
Statements that construction companies evaluate as weaknesses of BIM-VR integration and their correspondence.

- In post-construction processes, it is used for systematic facility management.
- In sales processes, it is possible to say that it is used to accelerate and increase sales, improve customer experience, provide new perspectives and receive feedback from customers.

As a result of scanning the websites of 10 selected construction companies, it is seen that the main centers of the construction companies examined are in the USA, for this reason, it is seen that the projects where BIM-VR technologies are used are mostly in the USA and that countries such as Canada, England, China and Japan, as well as the USA, are actively using these technologies. This can be explained by the fact that the USA, China, Japan and many European countries have introduced standards for the use of BIM. In the research conducted by McGraw-Hill in 2012, while the adoption rate of BIM in the construction sector was 28% in 2007, the adoption rate in 2012 increased to 71%. Again, in the research conducted by McGraw-Hill in 2010, the adoption rate of BIM use in European countries was stated as 36% [12, 33–34]. The

Firm	Structure Type	Country	BIM-VR Tools	BIM-VR Usage Purposes	BIM-VR Integration Strengths	BIM-VR Integration WEAKNESSES
Firm 1	Tunnel, Education, Health, Bridge, Housing, Airport, Office, Transportation, Shopping Mall	Czech Republic, Estonia, Canada, Denmark, Finland, Norway, Poland, Romania, Slovakia, the UK, the USA	Revit- Navisworks - UAS (Drone)	1. Accelerating Sales 2. Interactive Environments 3. Improving Customer Experience 4. Capturing New Perspectives 5. Delivering the Right Data to the Right Person at the Right Time on Site	1. Increased Security 2. Field Management 3. Strengthening Communication 4. Accuracy of Procurement Data 5. Reducing Waste 6. Reducing Costs 7. Increased Cooperation 8. Ease of Maintenance and Facility Management Decisions	1. Initial Costs 2. Adapting to Technology
Firm 2	Industry, Education, Health, Office	America, Korea, Singapore, Atlanta, Netherlands	Revit- BIM 360- Navisworks - Oculus Rift Headset	1. Project Tracking 2. Design Development 3. MEP Coordination 4. Security and Site Planning 5. Facility Management	1. Increasing the Participation of Project Stakeholders 2. Strengthening Communication 3. Analyzing the Challenges in the Project 4. Cost and Time Certainty 5. Occupational Safety Plans	1. Long-Term Work 2. Technology Sometimes Stays Only in the Office, Not Adapted to the Field

Firm	Structure Type	Country	BIM-VR Tools	BIM-VR Usage Purposes	BIM-VR Integration Strengths	BIM-VR Integration WEAKNESSES
Firm 3	Industry, Office, Mixed Use, Chemistry, Public	The USA	Laser scanning- Revit	1. Coordination of Building Systems 2. Improving Design 3. Project Visualization	1. Reducing Uncertainty 2. Predicting Conflicts 3. Finding Easy Solutions to Challenges 4. Simulation of Potential Impacts before Construction. 5. Flawless Management	1. Initial Costs 2. Training Employees on Technologies
Firm 4	Public, Housing, Office, Industry, Energy, Health, Mixed Use	Canada, England, Turkey, Iraq, Israel, Japan, Korea, Mexico, Spain, the USA	Revit - Navisworks - UAC (Drone) - Laser Scanning- VR Headset	1. Achieving Speed in Production 2. Project Visualization 3. Minimizing Problems During Construction	1. All Stakeholders Working in a Common Model 2. Ensuring Coordination between Teams 3. Ensuring Quality Controls 4. Ensuring Conflict Controls 5. Fast Field Logistics and Occupational Safety Planning	1. Long-term work 2. Human Resources Adapted to Technology

Firm	Structure Type	Country	BIM-VR Tools	BIM-VR Usage Purposes	BIM-VR Integration Strengths	BIM-VR Integration WEAKNESSES
Firm 5	Airport Cultural Center, Hotel, Industry, Public, Transportation	Kazakhstan, Italy, China, Philippines, the United States, India, Japan, Spain	HTC VIVE- Revit - Rhino - Grasshopper- Sketch Up- AutoCAD - Ms. Project -	1. Enabling the Presentation of the Design 2. A More Detailed Model 3. Seeing Details	1. Reduced Re-work 2. All Stakeholders Working on a Single Model 3. A Net Cost Amount 4. Impressive Models that Give a Sense of Depth	1. Initial Costs 2. Training Employees on Technologies
Firm 6	Energy, Education, Public, Mixed Use, Industry,	The USA, Canada	UAS (Unmanned Aerial Vehicles-Drone)- Laser Scanning - Revit - Navisworks	1. Mapping the Area for Excavation 2. Control of Conflicts	1. Rich 3D Maps 2. Ensuring Work Safety 3. Increased Coordination and Efficiency of Project Teams 4. Increased Customer Satisfaction 5. Ensuring Quality Control	1. Long-Term Work 2. Initial Costs
Firm 7	Education, Industrial, Public, Health, Healthcare, Interior	The USA	Revit- Laser scanning	1. Reducing risks 2. Impressive Design	1. Eliminating Risks 2. Seeing the Future of Designs 3. Easy Facility Management 4. Reducing Cost	1. Technological Investment

Firm	Structure Type	Country	BIM-VR Tools	BIM-VR Usage Purposes	BIM-VR Integration Strengths	BIM-VR Integration WEAKNESSES
Firm 8	Mixed Use, Education, Skyscraper, Public, Indoor, Transportation	The USA, Australia, Thailand, Netherlands	Unity Reflect - VIVE - Revit	1. Creating a sense of presence and depth 2. Performing Conflict Checks 3. Interactivity	1. Good Feedback 2. Savings 3. Eliminating Errors 4. Ease of Updating the Project	1. Technological Investment
Firm 9	Education, Health, Transportation, Office, Industry, Public	The USA, the United Kingdom	Revit - Rhino - BIM 360 - ArchiCAD - Tekla - Trimble - Navisworks - 3DsMAX	1. Safe Workplace 2. Centralizing Equipment Management 3. Coordination	1. Increasing Quality 2. Efficiency 3. Satisfaction 4. Eliminating Risks	1. Technological Investment 2. Providing Human Resources
Firm 10	Office, Airport, Interior, Cultural, Trade	The USA, the United Arab Emirates, China	BIM 360- HTC VİVE Headset	1. Sharing a Common Language 2. Making Effective Presentations to Customers	1. Saving Time 2. Preventing Excess Effort 3. Increasing Customer Communication 4. Being Complementary	1. Providing Human Resources

Table 5.
Reviewed construction companies and evaluation.

Strengths of the Use of BIM-VR Technology	Weaknesses of the Use of BIM-VR Technology
1. Increased Communication, Cooperation and Coordination	1. Start-up Costs
2. Cost and Project Time Certainty, Savings and Efficiency	2. Adaptation to Technology
3. Improving the Design with Details	3. Long-Term Work
4. Sharing a Common Language	4. Length of the Acclimatization Period
5. Seeing More Detailed Models and Details in 3D	5. Ensuring Human Resources Adapted to Technology
6. Ensuring Overlap and MEP Controls	6. Time Lost in Adaptation to Technology
7. Ease in Field Planning and Management	7. Cost and Duration of Trainings to be Provided to Employees
8. Ease of Work Safety and Risk Planning	8. Technology Stays Only in the Office and Failure to Integrate into the Field
9. Increase in Sales and Customer Satisfaction	9. Reluctance to Technology
10. Effective Presentation and Visualization to the Customer	

Table 6.
BIM-VR technology blog posts on the case studies on the websites of software and construction companies.

fact that BIM-VR integration is not actively seen in building production processes in Turkey is the lack of a national standard. To accelerate the widespread use of these technologies in the Turkish construction sector, national BIM standards and guidelines should be prepared through all relevant institutions, laws and regulations should be introduced and necessary incentives should be provided [35].

It is noteworthy that the working areas of construction companies and the types of buildings in which they use BIM-VR technologies are large projects such as education, health, public, industry, office and cultural centers, which include more complicated systems, where MEP controls should be done carefully and serve mass use.

When we look at the software and programs used by the examined construction companies, Revit is the most common BIM software, followed by Navisworks. When we look at the hardware and tools they use in VR technologies, it is possible to say that Headsets, Laser Scanners and Drones are used intensively.

According to the comments, interviews, blog posts and case studies examined throughout the study, the strengths and weaknesses of the use of BIM-VR integration in building production processes were inferred.

As a result of the study, if the strengths of the use of BIM-VR technologies in building production processes are interpreted systematically:

The most frequently mentioned strength of BIM-VR integration for project teams is the significant increase in communication, coordination and collaboration between stakeholders at all stages of the building production process.

With BIM and VR technology, it is important to be able to perform clash and MEP checks even at the design stage, to prevent errors arising from system selection before the construction process and to detect clashes that will cause changes in the construction phase, deficiencies or errors in the required volumes in advance. During the production phase, by performing 3D controls of overlap and MEP controls in the field, returns are reduced, and system controls can be performed. It also has benefits such as ease and accuracy in surveying during restoration projects.

The convenience of BIM and VR technologies in making site and risk planning in production, the construction of occupational safety and emergency action plans and the support of occupational safety trainings of the field team with simulations provide a lot of convenience in safety and site management stages. In the post-production process, BIM-VR integration facilitates documentation and renovation-repair plans in facility management.

The first of the strengths that BIM-VR integration provides to company performance is that all stakeholders share a common language in the design processes and work on a single model, accelerating the project processes, thus saving time at the design stage. The project budget is obtained clearly and precisely with the ability to clearly extract the cost of the project.

In addition, with details that can be viewed in 1/1 scale and in 3D, returns and reconstruction in production are reduced. In this way, cost and time loss will be prevented during the return, repair and reconstruction phases.

The strengths of BIM-VR integration for customers are that the design process can result in customer satisfaction by organizing multi-participant design meetings where the user (customer) can participate during the design phase of the project and see the 3D equivalent of any change request instantly. Customers can take an active role in the design process.

In addition, by observing the construction process live, they can be more interested and knowledgeable about the construction process.

In the sales stages, the project can be experienced in 3D with a sense of depth and presence even at the design stage of the project, and thanks to effective visualization methods, there is an increase in sales and customer satisfaction.

If the weaknesses of the use of BIM-VR technologies in building production processes are interpreted systematically:

The first weakness of BIM-VR integration seen by the project teams is the necessity to create a detailed model to integrate BIM-VR technologies into building production processes. Only by creating this detailed model, realistic visuals and MEP details can be accessed. To be able to handle each stage of the building production processes in a controlled and systematic manner, a detailed model must be entered into the system for an image with the expected detail. Compared to the traditional method in design processes, longer-term studies can be realized in the design phase of building production processes with BIM-VR integration. The reasons for this can be exemplified as detailed model creation, meetings with project stakeholders, analysis and conflict control tests. These studies carried out during the design phase are returned in the form of time and cost savings during the construction phase by identifying the problems that will be encountered during the construction process in advance.

The fact that the process of getting used to the technology is long and requires a process the project teams need to be supported with training and frequent audits.

Some user-related disruptions such as errors arising from file transfers and physiological problems are also considered weaknesses of BIM-VR technology. As we use these systems, the problems will be easily overcome.

The weakness of BIM-VR integration in terms of company performance is that the investment cost of BIM-VR technologies is somewhat high. It is true that the initial investment costs of BIM and VR technologies are somewhat high, but the use of these technologies will reduce the number of returns and reconstructions in the field, thus saving cost and time.

Another aspect of BIM-VR that is seen as a weakness is meeting the human resources to adapt to the technology. In addition, if a new team will not be established, the cost and duration of the training that needs to be given to current employees are also considered weaknesses in the transition to BIM-VR integration. The adaptation of employees to the technology, which is evaluated as a weakness, is part of the transition process to BIM-VR technology. There is no doubt that these weaknesses will turn into benefits with the time and savings gained in the following processes.

The weakness of BIM-VR integration by customers can be interpreted as reluctance and prejudice toward BIM-VR technology. Customers who think that using VR is dizzying and nauseating will realize that this weakness is actually a prejudice after their experience. This was observed in the company study.

According to the comments, interviews, blog posts and case studies analyzed throughout the study, potential opportunities and potential threats of the use of BIM-VR integration in building production processes were inferred. To interpret the potential opportunities of integrating BIM-VR technologies into the building production process:

- First, the savings that can be achieved in the use of resources draw attention as the biggest opportunity. With BIM-VR integration, all stages of the building production process will proceed in a planned and controlled manner and returns and modifications will be reduced. In this way, exhaustible resources such as water, materials and energy will be protected, and the wastage of these resources will be prevented.

- Another opportunity is the potential increase in labor quality. Having a project with all the details in the hands of field workers and being able to view these details in 3D and 1/1 scale, when necessary, will help more esthetic, smooth finish and error-free workmanship.

- Another opportunity is to train oneself as a human resource adapted to technology and to provide individual employment. It is an important opportunity to attract the attention of companies by training oneself to adapt to these technologies that have become a part of the construction industry globally, to become a sought-after expert and thus increase individual employment.

Possible threats of the use of BIM and VR technologies in the building production process:

- The requirement to train or have the personnel trained for adaptation to BIM and VR technologies. Since employers consider this as a waste of time and cost, they may seek human resources adapted to the technology. This situation may lead to a loss of employment in existing employees, albeit a small amount after digitalization. However, individuals who adapt themselves to these technologies can increase their individual employment.

- Another threat is the possibility that the design process may be slightly prolonged as everyone expresses his/her opinion with a very participatory approach in design meetings involving the client and the stakeholders of the project.

However, this will reduce the returns in the production process when the optimal design is achieved, saving time and cost.

- The last potential threat is that the technology may remain only in the office and not be used in the field, or that field teams may continue to use traditional methods due to lack of adequate equipment. In this case, it has become essential to invest in technology at the field level and to train field workers to adapt to technology.

6. Conclusion

The building production process is the most energy-consuming process in the global context, with the most intense loss of energy, time, cost and material due to mistakes. Many problems are encountered because of the use of traditional methods in building production processes. Some of the problems in the design stages can be listed as not being able to give a model to the client in meetings, drawing all projects such as installation, electricity, static on different platforms and not being able to provide integrity and not being able to access the project cost clearly. An important problem in cultural heritage buildings is inaccurate surveying. Problems in the construction stages can be given as noncompliant productions, clashes in MEP projects, the dependence of manufacturing controls on human, lack of attention to occupational safety or lack of training, lack of consideration of field logistics and organization. The problems in post-construction processes are the loss of the points paid attention during the construction phase during repair and renovation, difficulties in planning the renovation process. These problems constitute a very small percentage of the problems encountered in the sector [36–37]. Each revisit and reconstruction causes a loss of energy, time and material and eventually increased cost. The protracted building production processes with traditional methods have created the need and demand for a more systematic and organized building production process. In addition to all these, in the academic field, traditional education methods continue in the education of architecture and civil engineering students on building production processes, and while the sector continues to change in a global context, students are far away from developing technology and software.

The construction industry has always been open to innovation and adaptable to technology. Developing technology has found its reflection in building production processes, and BIM and VR technologies have become the preferred technologies in building production processes. The fact that most of the new generation VR tools work with BIM software has brought the two technologies together and helped them to become increasingly widespread in the sector. By including the integration of these two technologies in building production processes, the process progresses in a more optimized, more controlled and more systematic way. In the design phase of the use of building production processes in design, cultural heritage, construction and post-construction phases, the realism and sense of depth in visualization, the ability of all project stakeholders to work on a single model, the early detection of design flaws, the ability to experience analyses and clash tests in 3D and the ability to improve relationships and cooperation with customers are noteworthy. In cultural heritage buildings, the accuracy in taking surveys of the building, virtual environment museum at the end of restoration, can enable applications for educational and gaming purposes. With the use of BIM-VR in building production processes and increased communication and coordination between employees, project follow-up and manufacturing controls can be easily carried out, site organization and occupational safety planning can be easily provided and serious advantages to project management have become possible. Finally, in facility management in the post-construction process, there are many remarkable benefits such as planning for foreseen or unforeseen modifications and ease of documentation about the building. The integrated use of BIM and VR technologies from the design phase of the project to demolition provides a systematic, controlled and realistic management throughout the life cycle of the building. The integrated use of BIM-VR technologies throughout the life cycle of the building will provide the main benefits such as efficiency, savings and conservation of resources.

Research shows that the world has integrated BIM and VR tools in the last 5 to 10 years and BIM-VR integration has started to take its place in building production processes. This progress continues by rapidly adding new hardware to itself. In the studies examined, it has been observed that there is awareness of BIM and VR technologies; however, there is hesitation in the transition to these technologies, especially in developing countries. At this point, the lack of standards and guidelines to be taken as reference in the implementation and execution of project processes draws attention. In developed countries, the use of these technologies is mandatory in public tenders and included in contracts in the private sector.

As a result of this study, to take advantage of the benefits and potential opportunities of using BIM-VR technologies on a sectoral basis, developing countries should invest in this technology, revitalize the construction sector and support potential human resources with trainings on these technologies. Fairs, events and promotions should be organized for company owners, architects, engineers and even all sector employees to recognize and use BIM-VR integration, and conferences and workshops on the use of these technologies should be organized. The state's support and contribution to technological developments related to building production processes and granting privileges to those who use them will help us to see BIM-VR integration frequently in building production processes. Publications made on this subject should be shared with all sector stakeholders and a continuous flow of information from academia to the sector should be ensured. In the global context, it is evident that research on the benefits and use case of BIM-VR integration in a construction-oriented context such as design and constructability review, sequencing and work planning is limited and there is an information gap [21, 38]. It is necessary to support the use of different methods and case studies for academics and students who will continue their lives in academia in their studies on these technologies. In addition, in the academic field, the main actors in building production processes, such as architecture and engineering professions, should be supported to receive education by adapting to the developing technology and to use BIM and VR technologies during their university education. A thesis study on the use of VR in building science courses showed that Generation Z individuals are open to innovations and technology and contribute to the perception, comprehension and imagination of structural elements [39]. In addition, it should be ensured that the project actors in the building production processes with these systems in the sector are included in the training. In this way, it will be possible to train individuals who have a good command of current building production processes by providing sector and academic cooperation.

As suggestions for future studies on the subject, interviews can be conducted with construction companies that have transitioned to BIM-VR integration in the sector to contribute to the method. In these interviews, the study can be supported by asking about the benefits and damages they can identify in the construction production processes after the transition to BIM-VR integration. In addition, the purposes, benefits and damages of their use in design, construction, post-construction or cultural heritage projects can be determined through case studies in real projects where BIM-VR integration is used in building production processes.

In summary, the construction sector, which consumes the largest amount of resources, should be adapted to technology, innovations related to the sector should be followed, BIM-VR integration should be ensured in building production processes and potential opportunities related to BIM-VR integration should be transformed into real benefits. In addition to innovation on a sectoral basis, more analysis and case study-based publications on BIM-VR integration should be produced in academia. Only if the sector and academia come together and become innovative, the potentials of the construction industry can be accessed. With the continuity of the studies in this context, there is no doubt that BIM-VR technologies will find a place for themselves as a new traditional system in the construction sector of developing countries.

Acknowledgements

This book chapter is an extended version of a paper by the same authors, originally published in the *Turkish Journal of Civil Engineering*: Bozkurt G, Ersoy S, Yaşa E. Yapı Üretim Süreçlerinde BIM ve VR Uygulamalarının Entegrasyonunun Değerlendirilmesi. tjce. 2024;35(4):117-44, https://doi.org/10.18400/tjce.1244122. It includes additional data and analyses. Permission to use and extend the original work has been granted by the journal.

Conflict of interest

The authors declare no conflict of interest.

Author details

Gülce Bozkurt[1], Selahattin Ersoy[1] and Enes Yaşa[2*]

1 Istanbul University, Istanbul, Turkey

2 Istanbul Technical University, Istanbul, Turkey

*Address all correspondence to: enesyasa@itu.edu.tr

References

[1] Kaya E. Yapı üretim sürecinde yapım aşamasında kaliteyi etkileyen faktörler ve iş gücünün önemi [thesis]. İstanbul: İstanbul Teknik Üniversitesi; 1999

[2] Karabulut Ö. Yapı üretim sürecinde kaynak yönetimi karar alma modeli [thesis]. Ankara: Gazi Üniversitesi; 2007

[3] Yılmaz G, Güngör AA, Demirörs O. İnşaat sektöründe kullanılan bilgi ve iletişim teknolojileri. In: 10. Ulusal Yazılım Mühendisliği Sempozyumu (UYMS'xx16); 24–26 October 2016; Çanakkale. Çanakkale: UYMS; pp. 562-573

[4] Autodesk. BIM Advantage. Available from: https://www.autodesk.com.tr/solutions/bim/benefits-of-bim [Accessed: January 10, 2023]

[5] Petric J, Maver T, Conti G, Ucelli G. Virtual reality in the service of user participation in architecture. In: Proceedings of the CIB W78 Conference 2002—Distributing Knowledge in Building, 12–14 June; Aarhus. Aarhus: CIB W78; 2002. pp. 217-224

[6] Ofluoğlu S. Yapı Bilgi Modelleme: Yeni Nesil Mimari Yazılımlar. Mimar Sinan Üniversitesi, 2012. Available from: http://www.sayisalmimar.com/yayin/ybm.pdf [Accessed: February 1, 2023]

[7] Getuli V, Capone P, Bruttini A, Isaac S. BIM-based immersive virtual reality for construction workspace planning: A safety-oriented approach. Automation in Construction. 2020;**160**: 103160. DOI: 10.1016/j.autcon.2020.103160

[8] Halili A. An XML-based approach for geo-semantic data exchange from BIM to VR applications. Automation in Construction. 2021;**121**:103425. DOI: 10.1016/j.autcon.2020.103425

[9] Kim JI, Li S, Chen X, Keung C, Suh M, Kim TW. Evaluation framework for BIM-based VR applications in design phase. Journal of Computational Design and Engineering. 2021;**8**:910-922. DOI: 10.1093/jcde/qwab022

[10] Eastman C, Teicholz P, Sacks R, Liston K. BIM Handbook: A Guide to Building Information Modeling for Owners, Managers, Designers, Engineers and Contractors. Hoboken: Wiley; 2011. 633 p

[11] Yenigün İ, Yenigün K, Erdoğan S. Sanal Gerçekliğe Ticari Uygulama Yaklaşımları. GSI Journals Serie C: Advancements in Information Sciences and Technologies (AIST). 2020;**3**(1): 22-31

[12] Alizadehsalehi S, Hadavi A, Huang JC. From BIM to extended reality in AEC industry. Automation in Construction. 2020;**116**:103254. DOI: 10.1016/j.autcon.2020.103254

[13] HoloLens2/ Microsoft. Available from: https://www.microsoft.com/en-us/hololens [Accessed: January 5, 2024]

[14] RealWear. Available from: https://www.realwear.com/ [Accessed: January 5, 2024]

[15] Architecture, Engineering, Construction, and Operations. Unity. Available from: https://unity.com/solutions/architecture-engineering-construction [Accessed: January 4, 2024]

[16] V Unity, SHoP Architects: A Unity Reflect Case Study. SHop Architects. Available from: https://unity.com/case-

study/shop-architects#building-right-tool [Accessed: January 10, 2024]

[17] Kieferle J, Woessner U. BIM interactive—About combining BIM and virtual reality—A bidirectional interaction method for BIM models in different environments. In: Proceedings of the 33rd eCAADe Conference eCAADe; 16–18 September 2015; Vienna. Vienna: eCAADe; 2015. pp. 69-75

[18] Design Experience with BIM-VR Integration during the Design Phase MARSBIM. Available from: https://www.marsbim.com/blog/how-does-bim-advance-by-using-ar-and-vr-technology-in-construction/ [Accessed: June 26, 2023]

[19] Park H, Panya DS, Goo HN, Kim T, Seo J. BIM-based virtual reality and human behavior simulation for safety design. In: Proceedings of the 36th eCAADe Conference; 19–21 September 2018; Lodz. Lodz: eCAADe; 2018. pp. 823-832

[20] Moraru A, Pozanski K. Integrating BIM, virtual reality and serious gaming for effective collaboration and communication between end-users and the design team [thesis]. Aalborg: Aalborg Üniversitesi; 2020

[21] Johansson M, Roupé M. BIM and virtual reality (VR) at the construction site. In: 19th International Conference on Construction Applications of Virtual Reality (CONVR 2019); 13–15 November 2019; Bangkok. Bangkok: CONVR; pp. 1-10

[22] INCIDE. Available from: https://www.incide.it/en/the-construction-site-management-between-simulation-and-virtualization/ [Accessed: June 26, 2024]

[23] Muhammad AA, Yitmen İ, Alizadehsalehi S, Çelik T. Adoption of virtual reality (VR) for site layout optimization of construction projects. Teknik Dergi. 2020;**31**(2):9833-9850. DOI: 10.18400/tekderg.423448

[24] Afzal M, Shafiq MT. Evaluating 4D-BIM and VR for effective safety communication and training: A case study of multilingual construction job-site crew. Buildings. 2021;**11**:8. DOI: 10.3390/buildings11080319

[25] Field Training with BIM-VR Integration during the Building Production Phase, SERIOUSLAB. Available from: https://seriouslabs.com/ [Accessed: June 26, 2023]

[26] Şen G. AR/VR Destekli BIM teknolojileri ile tesis yönetimi. Yapı Bilgi Modelleme. 2021;**03**(01):12-22

[27] VDC / BIM. SKANSKA. Available from: https://www.usa.skanska.com/what-we-deliver/services/innovation/vdc–bim/ [Accessed: January 10, 2023]

[28] Azhar S. Role of visualization technologies in safety planning and management at construction jobsites. Procedia Engineering. 2017;**171**:215-226. DOI: 10.1016/j.proeng.2017.01.329

[29] Baik A. The use of interactive virtual BIM to boost virtual tourism in heritage sites, historic Jeddah. ISPRS International Journal of Geo-Information. 2021;**10**(9):577. DOI: 10.3390/ijgi10090577

[30] Zaker R, Coloma E. Virtual reality-integrated workflow in BIM-enabled projects collaboration and design review: A case study. Visualization in Engineering. 2018;**6**:4. DOI: 10.1186/s40327-018-0065-6

[31] Haggard KE. Case Study on Virtual Reality in Construction. 2017. Available from: https://digitalcommons.calpoly.edu/cgi/viewcontent.cgi?article=1041&context=cmsp [Accessed: February 1, 2023]

[32] Deniz G. Expanding applications of virtual reality in construction industry: A multiple case study approach. Journal of Construction Engineering, Management & Innovation. 2019;**2**(2):48-66. DOI: 10.31462/jcemi.2019.02048066

[33] Bernstein HM, Jones SA, Russo MA. The Business Value of BIM in North America: Multi-Year Trend Analysis and User Rating (2007–2012). Bedford: McGrawHill Construction; 2012. 72 p

[34] Bernstein HM, Jones SA, Gudgel JE. The Business Value of BIM in Europe, Getting Building Information Modeling to the Bottom Line in the United Kingdom, France and Germany. Bedford: McGrawHill Construction; 2010. 52 p

[35] Erdik M, Gökuç TY. Türk Yapı Sektöründe Yapı Bilgi Modellemesinin Adaptasyonu. BAUN Fen Bil. Enst. Dergisi. 2020;**22**(1):159-171. DOI: 10.25092/baunfbed.679588

[36] Proverbs DG, Holt GD, Cheok HY. Construction industry problems: The views of UK construction directors. In: 16th Annual ARCOM Conference; 6–8 September 2000; Glasgow. Manchester: ARCOM; 2000. pp. 73-81

[37] Jaffar N, Tharim AH, Shuib MN. Factors of conflict in construction industry: A literature review. In: 2nd International Building Control Conference 2011; 11–12 July 2011; Red Hook: Curran Associates, Inc.; 2011. pp. 193-202

[38] Akbay RB. Türk inşaat sektöründe yapı bilgi modellemesinin şantiyede kullanımına yönelik bir inceleme [thesis]. Istanbul: Fatih Sultan Mehmet Üniversitesi; 2021

[39] Özdoğan M. Mimarlık okullarında yapı bilgisi eğitiminin sanal gerçeklik sistemleri ile bütünleştirilmesi [thesis]. Konya: Konya Teknik Üniversitesi; 2021

Chapter 5

Research Trends and Innovative Technologies in Construction Conflict Management: A VOSviewer Analysis

Buse Un and Ercan Erdis

Abstract

Conflicts are common in construction projects, which are intrinsically complex and involve numerous stakeholders. If conflicts are not averted or settled, they may result in irrevocable losses for the parties involved. From past to present, researchers have concentrated on construction conflict management to comprehend conflicts and alleviate their adverse effects. This study aims to map the research regarding strategic management of construction conflict and quantitatively analyze of these investigations. In this context, studies selected according to certain criteria from the Web of Science database were analyzed using VOSviewer software. The analysis encompasses the countries producing these studies, co-authorship networks, publication years, citation metrics, the journals in which they were published, and the most utilized keywords. Moreover, an analysis of studies concerning the development of innovative technologies used in construction conflict management, including building information modeling, virtual reality and augmented reality, digital rehearsals, computer vision, and the Internet of Things sensors, was conducted. The study posits that the integration of these innovative technologies into the preliminary design and construction processes has the potential to avert the escalation of latent errors and risks into conflicts, claims, and disputes. In this regard, the study provides significant insights for both academics in the domain of construction project management and industry stakeholders.

Keywords: construction management, strategic management, construction industry, project management, construction conflict

1. Introduction

Construction projects are inherently prone to risks due to their involvement of numerous stakeholders and complex work cycles [1]. If risks in construction contracts are not allocated in a clear and balanced manner, they may lead to conflicts. If conflicts are not effectively managed, they can evolve into claims. If claims are not appropriately resolved, they may further escalate into disputes (**Figure 1**) [2, 3].

Figure 1.
The concepts of risk, conflict, claim, and dispute [2, 3].

In the construction industry, the concepts of conflict and dispute are often confused and regarded as almost identical. However, disputes are a result of conflicts. Conflict can be defined as differences in the objectives or goals of project participants [4]. In other words, conflicts occur when parties in a business relationship have differing perspectives and interests while striving to achieve a common goal [5]. Conflict is a process that can lead to either positive or negative outcomes. The resolution of this process can be achieved through coercion, collaboration, compromise, correction, negotiation, concession, or avoidance, depending on the personalities, education, and business culture of the involved parties, following the identification of the primary causes of the conflict.

Despite the fact that projects are getting more complicated, productivity-based labour management, material usage in accordance with technical specifications, technological tools, equipment, and machinery integration, effective financial management, and the agility of the management structure to adapt to continuous changes and technological developments in the sector could mitigate conflicts and, consequently, disputes. This is only possible if there are leaders and managers with emotional and cognitive intelligence who can recognize conflict in an organization. Furthermore, the managers are required to adopt and internalize the management functions of POCCC (planning, organization, coordination, command, and control). Otherwise, the parties could bring up claims, experience disputes, and apply dispute resolution methods [6].

Construction disputes, which can be resolved through mechanisms such as litigation, result in significant project delays and additional costs. Furthermore, different alternative dispute resolution (ADR) techniques, including negotiation, mediation, and early neutral evaluation, may be applied in construction disputes. However, this not only leads to reputational and financial losses for the parties involved but also causes projects to deviate from their original objectives [7]. Therefore, intervening in conflicts proactively is of paramount importance to prevent disputes.

Studies that conduct in-depth analyzes of construction conflicts and the development of new technological tools have made significant progress, aiming to minimize the damages caused by conflicts in the construction industry. The purpose of this chapter is to explore academic trends related to the strategic management of construction conflicts and innovative advancements that facilitate conflict resolution. To this end, an extensive analysis of pivotal studies in the field was initially undertaken, concentrating on research that investigates the causes and categories of construction conflicts to enhance comprehension of the topic. Subsequently, studies specifically centered on construction conflicts and innovative technologies were reviewed and evaluated. In the following phase, studies that met specific criteria were exclusively selected from the Web of Science database and quantitatively analyzed utilizing VOSviewer software. In this context, the publication trends of the selected articles were assessed, and the relevant journals as well as the most active countries conducting research in this area were identified. Additionally, the co-citation network within these studies was examined, and co-occurring keywords were identified.

The remainder of the study is structured as follows: Section 2 delineates the research methodology, whereas Section 3 focuses on the examination of construction conflicts. Section 4 evaluates pioneering technology-driven research on the topic. Section 5 provides quantitative analyzes of chosen studies. Section 6 highlights the principal findings, contributions, and limitations of the study.

2. Research methodology

This study analyzed research pertaining to conflict management within the construction industry. Initially, research analyzing construction conflicts and exploring innovative tools applicable to this domain was intently reviewed, and recommendations for future research and practitioners were provided. Moreover, this study conducted a bibliometric analysis of studies addressing conflicts and management strategies in the construction sector by utilizing the Web of Science (WoS) Core Collection. In this study, the WoS database was used because it is one of the world's leading citation databases [8]. This database includes articles from the most impactful journals worldwide, as well as conference proceedings and books. The scope of this research is limited to only the articles. The reason for this is that the quantity of articles on construction management and conflict surpasses that of other publication categories, providing comprehensive insights into the subject matter [9, 10]. In this vein, searches were performed on December 3, 2024, using the keywords

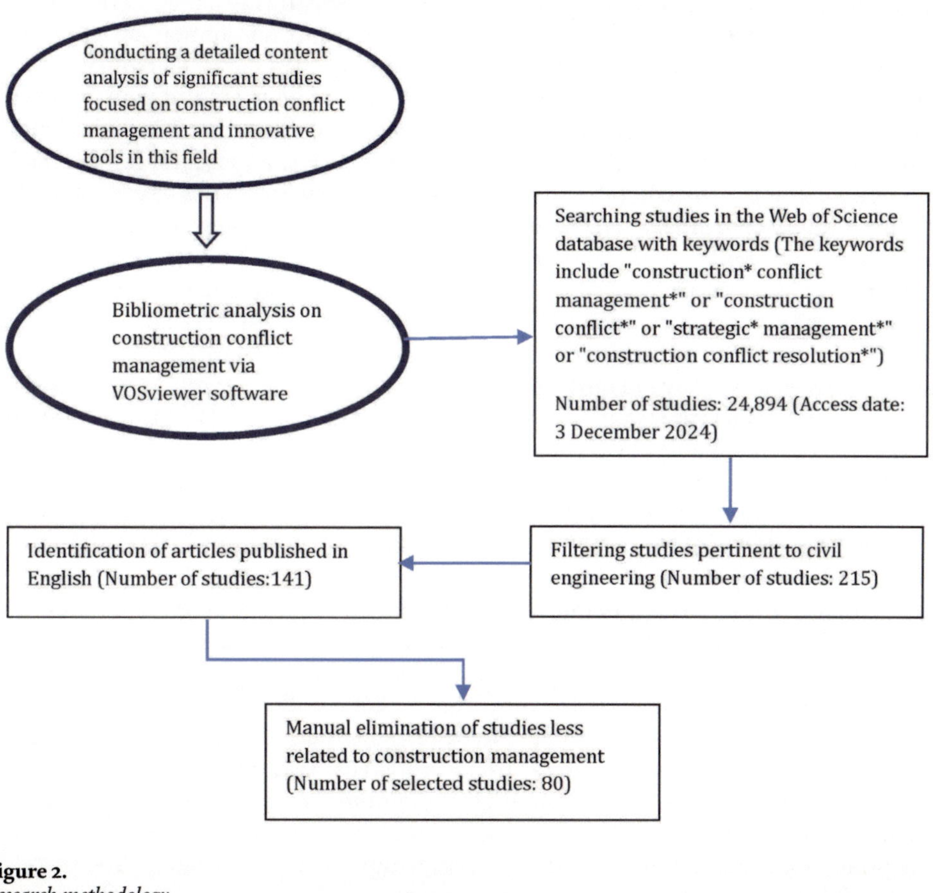

Figure 2.
Research methodology.

"construction* conflict management*" OR "construction conflict*" OR "strategic* management*" OR "construction conflict resolution*." As a result of this scan, 24,894 documents were reached. Subsequently, filtering was conducted on articles indexed in the Science Citation Index Expanded (SCI-Expanded) and Science Citation Index Expanded (SSCI) and published in English. No specific date range was selected for the related studies. After the exclusion of studies that were irrelevant to construction management, the remaining 80 studies were analyzed quantitatively in detail. This study performed bibliometric analyzes utilizing the free software VOSviewer version 1.6.20. The methodology of the present study is depicted in **Figure 2**.

3. Construction conflicts

Conflicts, much like in other industries, have a profound impact on the construction sector. Thus, many researchers have undertaken studies to gain a comprehensive understanding of the origins of risks, conflicts, and claims. Mahamid [11] examined the primary factors influencing conflicts between employers and contractors in his research. The study analyzed a survey conducted with project managers and an academic expert. Results of the analysis indicate that the primary factors contributing to conflicts include postponed monthly payments from owners, variation orders, poor work quality, rework, and deviations from the timeline for the project leading to delays in completion. In another investigation, Brockman [12] aimed to reveal the factors that trigger these conflicts and the cost of daily conflicts among individuals involved in construction sector projects. To achieve this, data obtained from individual interviews with a group of 74 people, consisting of workers and managers with at least 5 years of experience in the industry, was assessed with the critical incident method. According to the findings of the study, perceived safety issues, rework, lack of equipment/tools, construction planning, employer characteristics, communication deficiencies, and coordination between job roles are among the key factors that trigger interpersonal conflicts on construction sites. In addition, Vaux and Dority [13], according to the literature review, identified the factors that alleviate relationship conflicts in the construction sector. The study's results demonstrate that communication, conflict management strategies, and trust-building are the essential mitigating factors. Some scholars have identified the roots of construction conflicts by addressing contractual risks rather than interpersonal conflicts. Koc and Gurgun [14] identified risk factors that result in varying interpretations of contracts, potentially leading to conflicts between the parties involved. The authors compiled factors that could lead to conflicts regarding the reading of contracts into 18 categories and asked experts to assess the importance of these factors through a survey. The perceptions of the stakeholders who participated in the survey were analyzed and compared in four distinct groups: public-private sector, employer-contractor, employer-subcontractor, and contractor-subcontractor. According to the survey results, the majority of stakeholders identified the use of complex expressions in noun phrases and the improper use of references (e.g., it, they, their) as the most significant risk factors. The use of ambiguous words or phrases, as well as unnecessarily long sentences, were also recognized as other critical risk factors. The literature indicates that various types of conflicts will result in disputes. Cheung and Pang [15] sought to ascertain the principal factors contributing to the occurrence of construction disputes. The authors formulated a conceptual model delineating the structure of disputes, substantiated by literature and expert insights. Consequently, affective conflict and collaborative conflict are regarded as significant factors in this model. In addition to these, the strategic management of conflicts, along with their outcomes and impacts, has also been a focus of research. Min et al. [16] investigated the occurrence and characteristics of conflicts in public infrastructure projects. Their study proposes a six-part process framework to analyze and manage these conflicts, including conflict-causing factors, project characteristics, conflict events, government responses, main contractor responses, and conflict outcomes. Moreover, Karthikeyan and Manikandan [2] identified the causes and consequences of conflicts within the Indian construction sector in their research. A survey, developed with the assistance of literature, was administered to a group of employers, consultants, and contractors. The survey results indicate that the primary causes of conflicts are, the prioritization of objectives, alterations in site conditions, and interpersonal conflicts, respectively. The study also determined the primary effects of conflicts, including causing disputes, resulting in arbitration and time overruns.

In light of these studies, it was concluded that construction conflicts are affected by inadequate coordination and communication, interpersonal dynamics, site and project characteristics, planning, and contractual matters. Given that conflicts can result in irreparable harm to construction contracts, it is crucial to adopt an objective and strategic approach to conflict management through the application of new technologies.

4. Innovative technologies in construction conflict management

Research has highlighted that the implementation of technology-assisted proactive approaches can substantially enhance strategic conflict management in the construction industry. The technological tools generally attempt to mitigate the adverse impacts of conflicts on construction projects by enabling risk prediction and promoting amicable resolutions. In this context, studies focusing on decision support systems based on fuzzy logic or machine learning, BIM-integrated decision-making processes, and blockchain-based systems that preserved documents stand out [17]. A BIM-integrated conflict management approach was presented by Charehzehi et al. [18] to prevent construction conflicts from escalating into disputes. This study proposed a flowchart outlining the decision-making process supported by BIM in the event of a conflict. The study revealed that BIM functions, including clash detection, cost estimation, 4D planning, 3D visualization, and structural analysis, are effective in managing conflict factors. In addition, a system that generates, transmits, and synchronizes blocks through e-mail communication was designed by Kim et al. [19]. This system aims to preserve and manage documents generated throughout the construction project process securely. This research endorses the utilization of blockchain technology for the resolution of claims and disputes. Apart from these, a series of investigations [20–24] have attempted to predict the likelihood of disputes occurring in construction projects using various machine learning techniques. In these studies, researchers have generally focused on predicting the occurrence of disputes by considering variables such as project characteristics, case details, and the approaches of the parties involved. These studies are significant in terms of reporting the probability of conflicts escalating into disputes.

In addition to the systems proposed by academic research in this field, innovative technologies that have been employed in practice in recent years also contribute to conflict management in the construction industry. Integrating Industry 4.0

technologies or digital transformation technologies into the practical applications of strategic conflict management may be essential. In the report prepared by the Get It Right Initiative (GIRI) in 2024, the technologies to reduce errors in the design and construction process were expressed in a case study document, built principally around the categories under the headings of; checking technology (schedule and data checkers, checklists), automated generation technology (design configurators and automated scheduling systems), workflow engines, visualization systems (virtual reality and augmented reality, digital rehearsals), collaboration and communication tools, computer vision and Internet of Things sensors. In addition to the aforementioned best practice case studies, the report also covered digital technologies used in quality management systems, including digital twins and change management. Moreover, the construction industry already employs 3D laser scanners (3DLS), Radio Frequency Identification (RFID), wearable construction technology (WCT), big data (BD), autonomous networked robots (ANR), and nanotechnology (NT) [25, 26]. All this digital technology facilitates transparent and effective communication among stakeholders in both physical and virtual environments, enabling real-time interaction and data-driven dispute resolution through efficient collaboration, information, and document sharing. The use of up-to-date digital technologies in the pre-design and construction processes of construction projects can prevent potential hidden errors and risks arising from internal and external environments from turning into conflicts, claims, and disputes. Therefore, these digital technologies are vital in conflict management [26–28].

Researchers in the pertinent field assert that these developed technological tools will mitigate conflicts and facilitate project completion without deviating from their objectives. Undertaking more extensive research that illustrates the practical efficacy of these applications could be advantageous for industry professionals.

5. Results of the VOSviewer analysis

In this section, the results of the bibliometric analysis are presented with tables and network maps.

5.1 Publication year and citation trends of the studies

Figure 3 illustrates the annual total of published studies and their corresponding citations. In accordance with the constraints of this investigation, no publications were identified prior to 1994. It appears that the number of studies reached its maximum in 2018. The quantity of academic research on conflict management may have diminished post-2018 due to substantial events that have profoundly impacted social life, including the COVID-19 pandemic, financial crises, economic hardships, wars, and natural disasters.

5.2 The journals publishing the research

Table 1 displays the names of the journals in which the articles were published, along with the overall number of publications. Upon examination of the journals in which the studies were published, it was seen that three journals, Engineering Construction and Architectural Management, Journal of Management in Engineering, and Journal of Construction Engineering and Management, prominently contributed to the field.

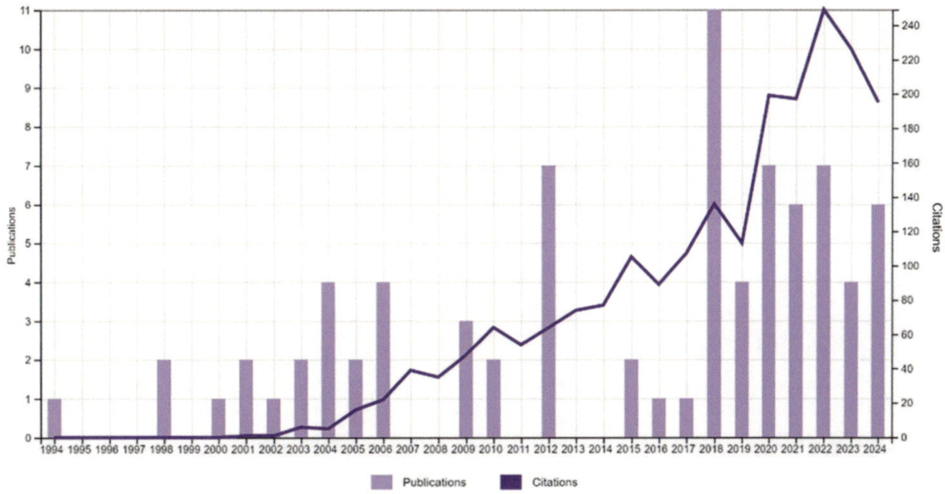

Figure 3.
Trends in the quantity of the articles and citations over the years [29].

Journals	No of articles
Automation in construction	3
Building and environment	1
Buildings	8
Civil engineering and environmental systems	1
Computer-aided civil and infrastructure engineering	1
Engineering construction and architectural management	22
Iranian journal of science and technology-Transactions of civil engineering	1
Journal of building engineering	1
Journal of civil engineering and management	1
Journal of construction engineering and management	11
Journal of construction engineering and management-Asce	4
Journal of management in engineering	20
Ksce journal of civil engineering	1
Structural safety	1
Teknik Dergi	1
Transportation finance, economics, and management	1
Transportation research record	2
Total	80

Table 1.
Journals in which articles are published [29].

5.3 Network of the co-occurrence keywords

In VOSviewer, the minimum number of occurrences of a keyword was set as 1, and a network related to co-occurring keywords was generated. The software detected 245

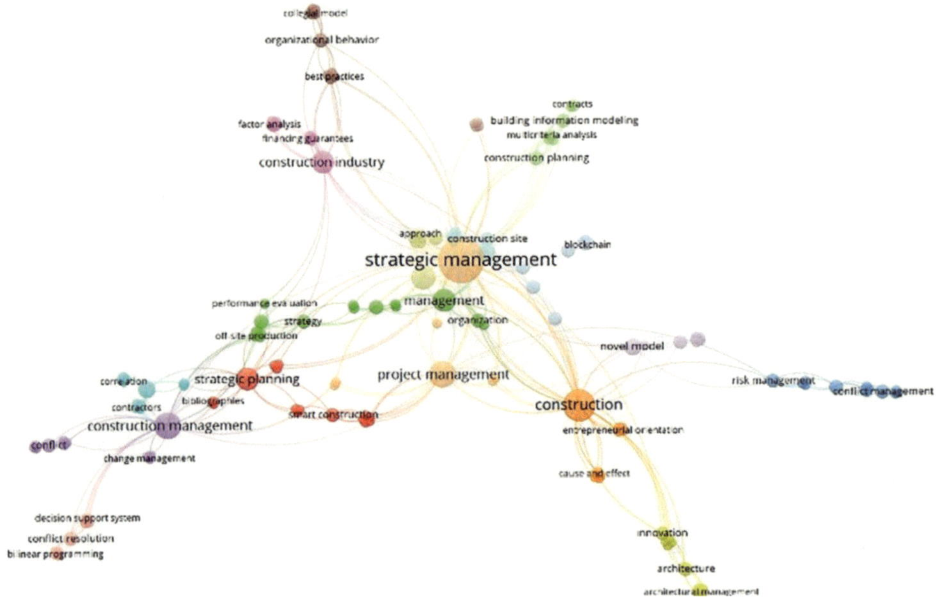

Figure 4.
Network visual of the co-occurring keywords [30].

keywords. **Figure 4** demonstrates the keyword clusters and links among them. The primary keywords include strategic management, construction, construction management, construction industry, and project management.

5.4 The most active countries in the studies

Table 2 illustrates a list of the most active countries in terms of publications. Furthermore, whereas the citations in the table indicate the number of citations for the articles, the total link strength represents the intensity of collaboration among

Country	Documents	Citations	Total link strength
China	22	799	10
USA	15	293	7
Australia	9	189	10
England	8	458	5
Canada	7	182	4
Taiwan	6	191	2
Türkiye	5	79	0
South Korea	4	115	4
Brazil	3	14	1
Chile	2	92	1

Table 2.
Analysis of the most active countries in publications and collaboration links by country [30].

Author	Documents	Citations	Total link strength
Zhang, G.	2	63	8
Anderson, S. D.	2	16	5
Arashpour, M.	1	40	5
Chan, A. P. C.	1	291	5
Chan, D. W. M.	1	291	5
Chan, E. H. W.	1	291	5
Chiang, Y. H.	1	291	5
Chinelli, C.K.	1	2	5
Guedes, A. l. A.	1	2	5
Gunarathna, C.	1	40	5
Haddad, A.N.	1	2	5
Ho, K. S. K	1	291	5
Jayasuriya, S.	1	40	5
Oladimeji, O.	1	2	5
Rodrigues, A. M.	1	2	5
Soares, C. A. P.	1	2	5
Tang, B. S.	1	291	5
Xue, X.	1	40	5
Yang, R. J.	1	40	5

Table 3.
Results of the analysis of co-authorship [30].

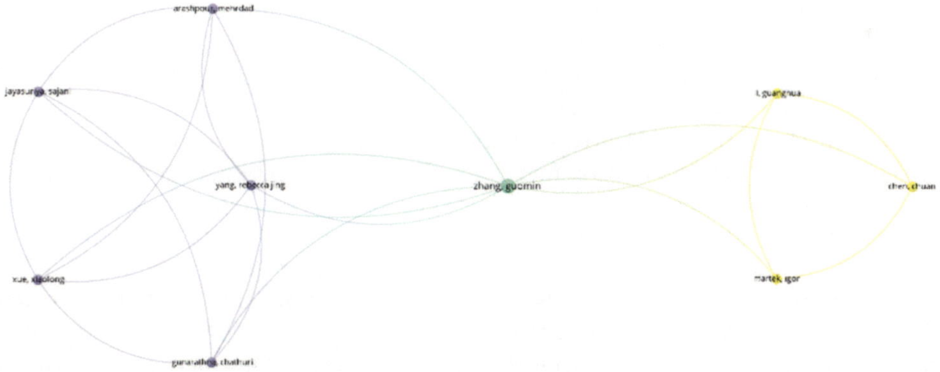

Figure 5.
Overlay visual of the co-authorship [30].

the countries. Here, China and the USA are the most active countries according to the number of documents. In terms of cooperation intensity, China and Australia are prominent. It is also noteworthy that UK-origin publications possess the second-highest citation count despite their limited number of publications.

5.5 Analysis of the co-authorship

Co-authorship analysis enables the review of collaborations among authors within the relevant domain. **Table 3** and **Figure 5** present the results of the co-authorship analysis. The nine authors with the most connections to each other are shown in **Figure 5**. The yellow sections denote authors who have produced relatively recent work.

6. Conclusions

The strategic management of conflicts that arise in the construction industry is essential for the successful completion of projects in terms of cost, time, and quality. Therefore, this study mapped conflict management studies within the construction sector and provided a comprehensive assessment of it. Furthermore, strategic and innovative advancements facilitating conflict management were examined. In this context, the relevant studies in the literature were first identified and analyzed to understand the essence of construction conflicts. After that, novel technological advancements targeted to mitigate the adverse impacts of conflicts were aggregated and assessed. Ultimately, the research based on the aforementioned criteria from the Web of Science database was examined utilizing VOSviewer software.

The examination's findings indicated that construction conflicts are influenced by site and project features, personal relationships, planning, contractual issues, and insufficient coordination and communication. It is crucial to conduct current research on the causes of conflicts in order to enable policymakers and construction professionals to implement preventative measures. In this context, the effects of economic, social, and technological changes on construction conflicts can be evaluated by consulting expert opinions.

Within the scope of this study, significant studies and practice applications of innovative technologies in construction conflict management have been examined. The research indicates that proactive approaches supported by technologies categorized as Industry 4.0 or digital transformation technologies can significantly improve conflict management in the construction industry. Technologies such as Building Information Modeling (BIM), wearable construction technology (WCT), digital twins (DT), virtual reality and augmented reality (AR), 3D laser scanners (3DLS), and the Internet of Things (IoT) can mitigate conflicts by enhancing construction management through real-time tracking of structural elements, site safety, and simulation. In addition, innovations such as Radio Frequency Identification (RFID), blockchain (BC), cloud computing platforms (CCP), and big data (BD) can enhance collaboration and transparency by actively participating in processes including supply chain management, customer relationship management, storage management systems, and enterprise resource planning. Finally, autonomous network robots (AND) and nanotechnology (NT) can contribute to quality and planning by improving existing materials and optimizing labour and equipment operations. These technologies can facilitate transparent communication between stakeholders, enabling real-time interaction and data-driven conflict resolution. The use of these technologies in the preliminary design and construction processes has the potential to prevent potential hidden errors and risks from escalating into conflicts, claims, and disputes.

In conclusion of the present study, it has been determined that these technological tools, which will enhance conflict management, should be disseminated and their efficacy in addressing conflicts should be documented. In this context, future research should concentrate on investigations that illustrate the practical application of developed technological tools for the benefit of industry professionals. Since it will shed light on problems in practice, case studies in this field are considered valuable. These technologies can mitigate conflicts during the project process; however, disagreements may also emerge among individuals concerning the utilization of these tools. The reviewed studies predominantly emphasize the introduction of technological tools via case analyzes while inadequately addressing the potential issues that may arise during their implementation. Considering that conflicts in the construction industry stem from factors such as project characteristics and planning, it is important to provide a detailed depiction through case studies of which tasks [31] (e.g., restoration, construction), which construction activities and which technologies (versions, tool models, etc.) [32] are used and how they are implemented. Furthermore, depending on the versions of the employed technologies, the work-related limitations imposed by these tools can be discerned, and these insights may be passed on to professionals actively engaged in the field. Moreover, by employing various case studies, a proactive strategy can be formulated to ascertain which contract types may yield advantageous results for specific parties in the event of such a conflict. All these efforts may clarify potential conflicts during the project process and aid professionals in parallel fields in acquiring insights and implementing preventive measures against issues. In this vein, integrating a comprehensive explanation of problem-resolution mechanisms within contracts can also mitigate the likelihood of a conflict-prone situation.

The validation of these employed technologies is also critically significant. Evaluating the outputs of a minimum of two technological tools for the same objective, aligned with the work's purpose, and comparing them with actual construction records can be instrumental in mitigating potential errors and conflicts.

Approaches can be developed to provide training to engineering students on how to use these technologies in practice. This will increase the academic performance of students [33], and the familiarity of graduates in the field with the use of current technologies will reduce possible conflicts. At this juncture, it is imperative to incorporate projects that encourage students to utilize these technologies within the curriculum and to create collaborations with industry, educational institutions, and government entities. Indeed, there is a necessity for research that encompasses a wider variety of parameters (including various technological tools and students from diverse geographies and cultures) to ascertain the impact of these applications on learning outcomes. In addition, to enhance conflict management, these tools can be improved with interdisciplinary collaborations by focusing on the missing or developing aspects of the tools used [34].

Upon examining the annual aggregate of articles published in reputable journals regarding the topic of conflict and their associated citations, it was determined that the quantity of publications and citations peaked in 2018, subsequently declining thereafter. The finding underscores the necessity for up-to-date studies, emphasizing the importance of raising both the number and quality of forthcoming research. The research findings indicate that the studies are published in esteemed journals that address construction conflict, are indexed in the Web of Science, and provide substantial contributions to the fields of construction and management. This finding is particularly significant for higher education students and academics planning to submit articles on conflict and conflict management since it provides guidance on which journals to target. The network of co-occurrence keywords analysis in VOSviewer reveals that the primary keywords in the articles are strategic management, construction, construction management, and project management. This is especially important for students, scholars, professionals, and individuals aiming to specialize in conflict and conflict management because it highlights the crucial terms and groups of words they should prioritize in their research within this domain. Moreover, the findings of the study demonstrate that China, the United States, and Australia are among the most active countries in strategic conflict management research. The study's results also showed the authors who collaborated the most in this field. For those working in this field of research and seeking careers, this outcome matters since it shows the countries that generate the most cited and popular research, as well as the authors who are eager to collaborate.

This study provides significant contributions by introducing management tools that construction professionals can integrate into practice, guiding their conflict management strategies. Additionally, this section maps out the body of work in construction conflict management, offering valuable insights to researchers intending to study or already working in this field.

The present investigation has certain limitations. This study exclusively examined articles indexed in the WoS database for quantitative analysis. Subsequent research may examine various database types and publications, comparing the findings of this study. Additionally, this review includes studies published up to 2024 in Civil Engineering. Further investigations can be updated in accordance with new research in this field.

Acknowledgements

Thanks to The Scientific Tecnological Research Council of Türkiye (TUBITAK-2211/A General Domestic Ph.D. Scholarship Program) and The Council of Higher Education (100/2000 Ph.D. scholarship program) for B. UN.

Conflict of interest

The authors declare no conflict of interest.

Author details

Buse Un[1*] and Ercan Erdis[2]

1 Department of Civil Engineering, Iskenderun Technical University, Hatay, Türkiye

2 Department of Architecture, Iskenderun Technical University, Hatay, Türkiye

*Address all correspondence to: buseun.mfbe19@iste.edu.tr

References

[1] Kalogeraki M, Antoniou F. Claim management and dispute resolution in the construction industry: Current research trends using novel technologies. Buildings. 2024;**14**:967. DOI: 10.3390/buildings14040967

[2] Karthikeyan R, Manikandan T. A study on causes and effects of conflicts in Indian construction projects. International Research Journal of Engineering and Technology (IRJET). 2017;**04**:1153-1170

[3] Wang P. The aetiology and progression of construction disputes between client and contractor in the UK [thesis]. Manchester: The University of Manchester; 2019

[4] Cheung SO, Yiu TW, Yeung SF. A study of styles and outcomes in construction dispute negotiation. Journal of Construction Engineering and Management. 2006;**132**:805-814. DOI: 10.1061/(asce)0733-9364(2006)132:8(805)

[5] Alshahrani S. Development the a dispute resolution framework to improve the efficiency of dispute resolution in Saudi construction projects [thesis]. Manchester: The University of Salford; 2017

[6] İlban T. İnşaat sektöründe çatışma yönetiminin kültürel boyutu[thesis]. Istanbul: Istanbul Technical University; 2008

[7] Chong HY, Zin RM. Selection of dispute resolution methods: Factor analysis approach. Engineering Construction and Architectural Management. 2012;**19**:428-443. DOI: 10.1108/09699981211237120

[8] Clarivate. Web of Science Core Collection Overview [Internet]. 2025. Available from: https://webofscience.help.clarivate.com/en-us/Content/wos-core-collection/wos-core-collection.htm [Accessed: February 02, 2025]

[9] Liu N, Zhou R, Jin R, Xiao Q, Hu Z. Scientometric review of construction conflict from 1991 to 2020. Engineering Construction and Architectural Management. 2023;**30**:4210-4228. DOI: 10.1108/ECAM-06-2021-0535

[10] Zheng X, Le Y, Chan APC, Hu Y, Li Y. Review of the application of social network analysis (SNA) in construction project management research. International Journal of Project Management. 2016;**34**:1214-1225. DOI: 10.1016/j.ijproman.2016.06.005

[11] Mahamid I. Effect of conflicts on the contracting business failure in the construction industry. Journal of King Saud University - Engineering Sciences. 2024;**36**:498-503. DOI: 10.1016/j.jksues.2023.12.003

[12] Brockman JL. Interpersonal conflict in construction: Cost, cause, and consequence. Journal of Construction Engineering and Management. 2014;**140**:04013050. DOI: 10.1061/(asce)co.1943-7862.0000805

[13] Vaux JS, Dority B. Relationship conflict in construction: A literature review. Conflict Resolution Quarterly. 2020;**38**:47-72. DOI: 10.1002/crq.21286

[14] Koc K, Gurgun AP. Assessment of readability risks in contracts causing conflicts in construction projects. Journal of Construction Engineering and Management.

2021;**147**:04021041. DOI: 10.1061/(asce) co.1943-7862.0002050

[15] Cheung SO, Pang KHY. Anatomy of construction disputes. Journal of Construction Engineering and Management. 2013;**139**:15-23. DOI: 10.1061/(asce)co.1943-7862. 0000532

[16] Min JH, Jang W, Han SH, Kim D, Kwak YH. How conflict occurs and what causes conflict: Conflict analysis framework for public infrastructure projects. Journal of Management in Engineering. 2018;**34**:04018019. DOI: 10.1061/(asce) me.1943-5479.0000625

[17] Zeberga MS, Haaskjold H, Hussein B. Digital technologies for preventing, mitigating, and resolving contractual disagreements in the aec industry: A systematic literature review. Journal of Construction Engineering and Management. 2024;**150**:3124002. DOI: 10.1061/JCEMD4.COENG-14032

[18] Charehzehi A, Chai CS, Md Yusof A, Chong HY, Loo SC. Building information modeling in construction conflict management. International Journal of Engineering Business Management. 2017;**9**:1-18. DOI: 10.1177/1847979017746257

[19] Kim EW, Park MS, Kim K, Kim KJ. Blockchain-based automatic tracking and extracting construction document for claim and dispute support. KSCE Journal of Civil Engineering. 2022;**26**:3707-3724. DOI: 10.1007/s12205-022-2181-z

[20] Ayhan M, Dikmen I, Birgonul MT. Predicting the occurrence of construction disputes using machine learning techniques. Journal of Construction Engineering and Management. 2021;**147**:04021022. DOI: 10.1061/(asce) co.1943-7862.0002027

[21] Chou J-S, Cheng M-Y, Wu Y-W, Pham A-D. Optimizing parameters of support vector machine using fast messy genetic algorithm for dispute classification. Expert Systems with Applications. 2014;**41**:3955-3964. DOI: 10.1016/j. eswa.2013.12.035

[22] Chou JS, Hsu SC, Lin C-W, Chang Y-C. Classifying influential for project information to discover rule sets for project disputes and possible resolutions. International Journal of Project Management. 2016;**34**:1706-1716. DOI: 10.1016/j.ijproman.2016.10.001

[23] Chou JS, Tsai CF, Lu YH. Project dispute prediction by hybrid machine learning techniques. Journal of Civil Engineering and Management. 2013;**19**:505-517. DOI: 10.3846/13923730.2013.768544

[24] Chou J-S, Lin C. Predicting disputes in public-private partnership projects: Classification and ensemble models. Journal of Computing in Civil Engineering. 2013;**27**:51-60. DOI: 10.1061/(asce) cp.1943-5487.0000197

[25] Aladağ H. Türk inşaat sektöründe dijital dönüşüm uygulamaları üzerine bir araştırma. Mühendislik Bilimleri ve Tasarım Dergisi. 2022;**10**:973-986. DOI: 10.21923/jesd.1085516

[26] Get It Right Initative. The Use of Technology to Reduce Errors in Design and Construction: Best Practice Casebook. London, England: Research Report; 2024

[27] Zhao X, Zhang M, Fan X, Sun Z, Li M, Li W, et al. Extended reality for safe and effective construction management: State-of-the-art, challenges, and future directions. Buildings. 2023;**13**. DOI: 10.3390/ buildings13010155

[28] Khan A, Sepasgozar S, Liu T, Yu R. Integration of bim and immersive technologies for AEC: A scientometric-SWOT analysis and critical content review. Buildings. 2021;**11**:1-126. DOI: 10.3390/buildings11030126

[29] Clarivate. Web of Science [Internet]. 2025. Available from: https://mjl.clarivate.com/home [Accessed: February 02, 2025]

[30] Van Eck NJ, Waltman L. VOSviewer (Version 1.6.20) [Internet]. 2023. Available from: https://www.vosviewer.com [Accessed: January 1, 2025]

[31] Pavelka K, Pavelka K, Běloch L. A reconstruction of the shrine of the prophet nahum: An analysis of 3D documentation methods and data transfer technology for virtual and augmented realities. Applied Sciences. 2025;**15**. DOI: 10.3390/app15021000

[32] Park S, Menassa C, Kamat V. Integrating large language models with multimodal virtual reality interfaces to support collaborative human-robot construction work. ArXivOrg. 2024;**39**:1-39. DOI: 10.1061/JCCEE5.CPENG-6106

[33] Shrestha S, Shan Y, Emerson R, Hosseini Z. Developing and usability testing of an augmented reality tool for online engineering education. IEEE Transactions on Learning Technologies. 2025;**18**:13-24. DOI: 10.1109/TLT.2024.3520413

[34] Toyin JO, Sattineni A, Wetzel EM, Fasoyinu AA, Kim J. Augmented reality in U.S. construction: Trends and future directions. Automation in Construction. 2025;**170**:105895. DOI: 10.1016/j.autcon.2024.105895